The Cryptoclub

Using Mathematics to Make and Break Secret Codes

Workbook

Janet Beissinger
Vera Pless

CRC Press
Taylor & Francis Group
Boca Raton London New York

CRC Press is an imprint of the
Taylor & Francis Group, an **informa** business

AN A K PETERS BOOK

CRC Press
Taylor & Francis Group
6000 Broken Sound Parkway NW, Suite 300
Boca Raton, FL 33487-2742

First issued in hardback 2017

CRC Press is an imprint of Taylor & Francis Group, an Informa business

No claim to original U.S. Government works

ISBN 13: 978-1-138-41314-6 (hbk)
ISBN 13: 978-1-56881-298-4 (pbk)

This book contains information obtained from authentic and highly regarded sources. Reasonable efforts have been made to publish reliable data and information, but the author and publisher cannot assume responsibility for the validity of all materials or the consequences of their use. The authors and publishers have attempted to trace the copyright holders of all material reproduced in this publication and apologize to copyright holders if permission to publish in this form has not been obtained. If any copyright material has not been acknowledged please write and let us know so we may rectify in any future reprint.

Except as permitted under U.S. Copyright Law, no part of this book may be reprinted, reproduced, transmitted, or utilized in any form by any electronic, mechanical, or other means, now known or hereafter invented, including photocopying, microfilming, and recording, or in any information storage or retrieval system, without written permission from the publishers.

Trademark Notice: Product or corporate names may be trademarks or registered trademarks, and are used only for identification and explanation without intent to infringe.

**Visit the Taylor & Francis Web site at
http://www.taylorandfrancis.com**

**and the CRC Press Web site at
http://www.crcpress.com**

 This material is based upon work supported by the National Science Foundation under Grant No. 0099220. Any opinions, findings, and conclusions or recommendations expressed in this material are those of the author(s) and do not necessarily reflect the views of the National Science Foundation.

The Library of Congress has cataloged the book associated with this workbook as follows:

Beissinger, Janet.
  The cryptoclub : using mathematics to make and break secret codes / Janet Beissinger, Vera Pless.
    p. cm.
  ISBN-13: 978-1-56881-223-6 (alk. paper)
  ISBN-10: 1-56881-223-X (alk. paper)
  1. Mathematics--Juvenile literature. 2. Cryptography--Juvenile literature.
  I. Pless, Vera. II. Title.

  QA40.5.B45 2006
  510--dc22   2006002743

Contents

Chapter 1: Caesar Ciphers
(Text page 4)

a	b	c	d	e	f	g	h	i	j	k	l	m	n	o	p	q	r	s	t	u	v	w	x	y	z
D	E	F	G	H	I	J	K	L	M	N	O	P	Q	R	S	T	U	V	W	X	Y	Z	A	B	C

Caesar cipher with shift of 3

1. a. Encrypt "keep this secret" with a shift of 3.

plaintext:	k	e	e	p		t	h	i	s		s	e	c	r	e	t	
ciphertext:																	

   b. Encrypt your teacher's name with a shift of 3.

plaintext:															
ciphertext:															

2. Decrypt the answers to the following riddles. They were encrypted using a Caesar cipher with a shift of 3.

   a. **Riddle:** What do you call a sleeping bull?
   Answer:

plaintext:											
ciphertext:	D		E	X	O	O	G	R	C	H	U

   b. **Riddle:** What's the difference between a teacher and a train?
   Answer:

plaintext:													
ciphertext:	W	K	H		W	H	D	F	K	H	U		V	D	B	V

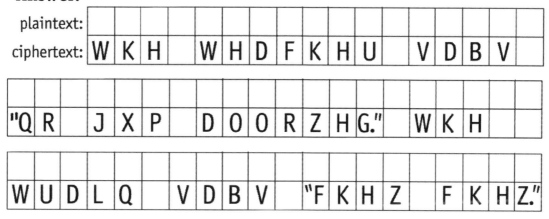

"Q	R		J	X	P		D	O	O	R	Z	H G."		W	K	H

W	U	D	L	Q		V	D	B	V		"F	K	H	Z		F	K	H Z."

(Text page 5)

a	b	c	d	e	f	g	h	i	j	k	l	m	n	o	p	q	r	s	t	u	v	w	x	y	z
E	F	G	H	I	J	K	L	M	N	O	P	Q	R	S	T	U	V	W	X	Y	Z	A	B	C	D

Caesar cipher with shift of 4

3. Decrypt the following note Evie wrote to Abby. She used a Caesar cipher with a shift of 4 like the one above.

plaintext:
ciphertext: | W | S | V | V | C. | | P | I | X'W | | Y | W | I | | |

| G | M | T | L | I | V | W | | J | V | S | Q | | R | S | A | | S | R. |

4. Use a shift of 3 or 4 to encrypt someone's name. It could be someone in your class or school or someone your class has learned about. (You'll use this to play Cipher Tag.)

plaintext:
ciphertext:

© 2006 A K Peters, Ltd., Wellesley, MA

The Cryptoclub: Using Mathematics to Make and Break Secret Codes

(Text pages 6–7)

5. a. Encrypt "private information" using a cipher wheel with a shift of 5. (Shift the inner wheel five letters counterclockwise.)

plaintext:	p	r	i	v	a	t	e		i	n	f	o	r	m	a	t	i	o	n
ciphertext:																			

b. Encrypt your school's name using a cipher wheel with a shift of 8.

plaintext:																			
ciphertext:																			

Use your cipher wheel to decrypt the answers to the following riddles:

6. **Riddle:** What do you call a dog at the beach?
   Answer (shifted 4):

plaintext:																			
ciphertext:	E		L	S	X		H	S	K										

7. **Riddle:** Three birds were sitting on a fence. A hunter shot one. How many were left?
   Answer (shifted 8):

plaintext:																			
ciphertext:	V	W	V	M.		B	P	M		W	B	P	M	Z	A				

N	T	M	E		I	E	I	G.											

Name _____ **Date**_____

(Text page 7)

8. **Riddle:** What animal keeps the best time?
    Answer (shifted 10):

plaintext:

ciphertext:
K		G	K	D	M	R	N	Y	Q								

9. Write your own riddle and encrypt the answer. Put your riddle on the board or on a sheet of paper that can be shared with the class later on. (Tell the shift.)

Riddle: _____

_____

_____

_____

Answer:

plaintext:

ciphertext:

© 2006 A K Peters, Ltd., Wellesley, MA

The Cryptoclub: Using Mathematics to Make and Break Secret Codes

Chapter 1: Caesar Ciphers

Name _____ Date _____

Chapter 2: Sending Messages with Numbers
(Text page 10)

a	b	c	d	e	f	g	h	i	j	k	l	m	n	o	p	q	r	s	t	u	v	w	x	y	z
0	1	2	3	4	5	6	7	8	9	10	11	12	13	14	15	16	17	18	19	20	21	22	23	24	25

1. a. **Riddle:** What kind of cookies do birds like?
   Answer:

plaintext:														
ciphertext:	2	7	14	2	14	11	0	19	4		2	7	8	17	15

   b. **Riddle:** What always ends everything?
   Answer:

plaintext:											
ciphertext:	19	7	4		11	4	19	19	4	17		6

Return to Text

2. a. Encrypt using the cipher strip at the top of the page.

plaintext:	J	a	m	e	s		B	o	n	d							
ciphertext:																	

   b. Encrypt using this cipher strip that is shifted 3.

a	b	c	d	e	f	g	h	i	j	k	l	m	n	o	p	q	r	s	t	u	v	w	x	y	z
3	4	5	6	7	8	9	10	11	12	13	14	15	16	17	18	19	20	21	22	23	24	25	0	1	2

plaintext:	J	a	m	e	s		B	o	n	d							
ciphertext:																	

   c. Describe how you can use arithmetic to get your answer to 2b from your answer to 2a.

   _____

   _____

   _____

© 2006 A K Peters, Ltd., Wellesley, MA

The Cryptoclub: Using Mathematics to Make and Break Secret Codes

(Text page 11)

3. Encrypt the following with the given shift:

a. shift 4

plaintext:	L	i	n	c	o	l	n
numbers:							
shifted numbers:							

b. shift 5

plaintext:	L	u	k	e
numbers:				
shifted numbers:				

c. shift 3 (What is different about encrypting the letter **x**?)

plaintext:	e	x	p	e	r	i	m	e	n	t
numbers:										
shifted numbers:										

_____

_____

_____

Return to Text

4. What numbers between 0 and 25 are equivalent on the circle to the following numbers?

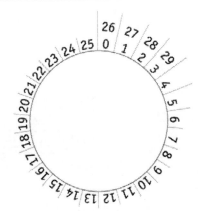

a. 28 _____   b. 29 _____

c. 30 _____   d. 34 _____

e. 36 _____   f. 52 _____

5. Describe an arithmetic pattern that tells how to match a number greater than 25 with an equivalent number between 0 and 25.

_____

_____

_____

6. Encrypt each word by adding the given amount. Your numbers should end up between 0 and 25.

a. add 4

plaintext:	x	-	r	a	y
numbers:					
shifted numbers:					

b. add 10

c	r	y	p	t	o	g	r	a	p	h	y

© 2006 A K Peters, Ltd., Wellesley, MA

The Cryptoclub: Using Mathematics to Make and Break Secret Codes

(Text page 12)

a	b	c	d	e	f	g	h	i	j	k	l	m	n	o	p	q	r	s	t	u	v	w	x	y	z
0	1	2	3	4	5	6	7	8	9	10	11	12	13	14	15	16	17	18	19	20	21	22	23	24	25

Cipher strip (no shift)

7. Jenny encrypted this name by adding 3. Decrypt to find the name.

↑ plaintext:

numbers:

shifted numbers: 14 | 11 | 14 | 3 | 10

8. **Riddle:** Why doesn't a bike stand up by itself?
   Answer (encrypted by adding 3):

↑ plaintext:

numbers:

shifted numbers: 11 | 22 | ' | 21 | | 22 | 25 | 17 | | 22 | 11 | 20 | 7 | 6

9. **Riddle:** What do you call a monkey who loves to eat potato chips?
   Answer (encrypted by adding 5):

↑ plaintext:

numbers:

shifted numbers: 5 | | 7 | 12 | 13 | 20 | | 17 | 19 | 18 | 15

10. **Riddle:** What is a witch's favorite subject?
    Answer (encrypted by adding 7):

↑ plaintext:

numbers:

shifted numbers: 25 | 22 | 11 | 18 | 18 | 15 | 20 | 13

11. **Challenge.** This is a name that was encrypted by adding 3.
    a. Decrypt by subtracting.

↑ plaintext:

numbers:

shifted numbers: 22 | 11 | 15 | 15 | 1

    b. What happens to the 1? What can you do to fix the problem?

_____

_____

© 2006 A K Peters, Ltd., Wellesley, MA

The Cryptoclub: Using Mathematics to Make and Break Secret Codes

(Text page 13)

12. What numbers between 0 and 25 are equivalent on the circle to the following numbers?

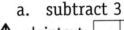

    a. 26 _____   b. 28 _____   c. −1 _____

    d. −2 _____   e. −4 _____   f. −10 _____

13. Describe an arithmetic pattern that tells how to match a number less than 0 with an equivalent number between 0 and 25.

_____

_____

_____

14. Decrypt by subtracting. Replace negative numbers with equivalent numbers between 0 and 25.

    a. subtract 3

plaintext:

numbers:

shifted numbers: | 18 | 11 | 2 | 2 | 3 |

    b. subtract 10

plaintext:

numbers:

shifted numbers: | 3 | 10 | 7 | 18 |

    c. subtract 15

plaintext:

numbers:

shifted numbers: | 7 | 4 | 13 |

15. **Riddle:** What do you call a chair that plays guitar?

Answer (encrypted by adding 10):

plaintext:

numbers:

shifted numbers: | 10 | | 1 | 24 | 12 | 20 | 14 | 1 | | | | | | | | |

16. **Riddle:** How do you make a witch itch?

Answer (encrypted by adding 20):

plaintext:

numbers:

shifted numbers: | 13 | 20 | 4 | 24 | | 20 | 16 | 20 | 18 | | 1 | 24 | 11 | | 16 | |

© 2006 A K Peters, Ltd., Wellesley, MA

The Cryptoclub: Using Mathematics to Make and Break Secret Codes

© 2006 A K Peters, Ltd., Wellesley, MA

Name _____ Date_____

(Text page 16)

17. a. To decrypt the riddle in Question 15, you could subtract 10. What number could you add to get the same answer as subtracting 10?

_____

   b. Here is the answer to the riddle in Question 15. Decrypt it again, adding or subtracting as necessary to avoid negative numbers and numbers greater than 25.

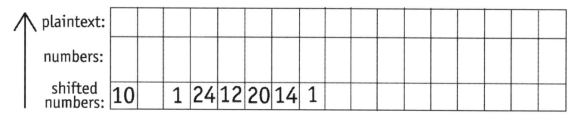

plaintext:																				
numbers:																				
shifted numbers:	10		1	24	12	20	14	1												

18. a. Suppose that you encrypted a message by adding 9. Tell two different ways you could decrypt it.

_____

   b. This message was encrypted by adding 9. Decrypt by adding or subtracting to avoid negative numbers and numbers greater than 25.

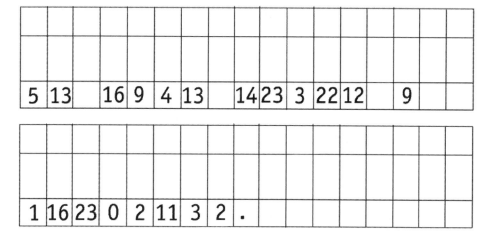

5	13		16	9	4	13		14	23	3	22	12		9	

1	16	23	0	2	11	3	2	.							

The Cryptoclub: Using Mathematics to Make and Break Secret Codes

(Text page 16)

19. a. Suppose that you encrypted a message by adding 5. Tell two different ways you could decrypt it.

_____

_____

b. In general, suppose that you encrypted a message by adding an amount n. Tell two different ways you could decrypt it.

_____

_____

For Questions 20–23, add or subtract as necessary to make your calculations simplest.

20. **Riddle:** Imagine that you're trapped in a haunted house with a ghost chasing you. What should you do?
   Answer (encrypted by adding 10):

plaintext:																				
numbers:																				
shifted numbers:	2	3	24	25		18	22	10	16	18	23	18	23	16						

21. **Riddle:** Why must a doctor control his temper?
   Answer (encrypted by adding 11):

plaintext:																				
numbers:																				
shifted numbers:	12	15	13	11	5	3	15		18	15		14	25	15	3	24	'	4		

7	11	24	4		4	25		22	25	3	15		18	19	3		0	11	4	19	15	24	4	3

© 2006 A K Peters, Ltd., Wellesley, MA

The Cryptoclub: Using Mathematics to Make and Break Secret Codes

(Text page 17)

22. **Riddle:** What is the meaning of the word "coincide"?
    Answer (encrypted by adding 7):

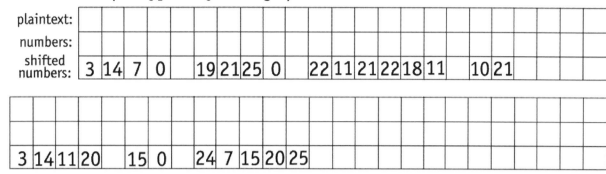

plaintext:																	
numbers:																	
shifted numbers:	3	14	7	0		19	21	25	0		22	11	21	22	18	11	10	21

3	14	11	20		15	0		24	7	15	20	25

23. Abby was learning about life on the frontier. "Peter," she said, "Where is the frontier?" Decrypt Peter's reply (encrypted by adding 13).

plaintext:																				
numbers:																				
shifted numbers:	6	20	13	6	'	5		13		5	21	24	24	11		3	7	17	5	6	21	1	0.

11	1	7		1	0	24	11		20	13	8	17		13		24	17	18	6		17	13	4

13	0	16		13		4	21	19	20	6		17	13	4	.

© 2006 A K Peters, Ltd., Wellesley, MA

The Cryptoclub: Using Mathematics to Make and Break Secret Codes

Here are some blank tables for you to make your own messages.

© 2006 A K Peters, Ltd., Wellesley, MA

The Cryptoclub: Using Mathematics to Make and Break Secret Codes

© 2006 A K Peters, Ltd., Wellesley, MA

Name _____ **Date**_____

Chapter 3: Breaking Caesar Ciphers
(Text page 21)

1. Decrypt Dan's note to Tim.

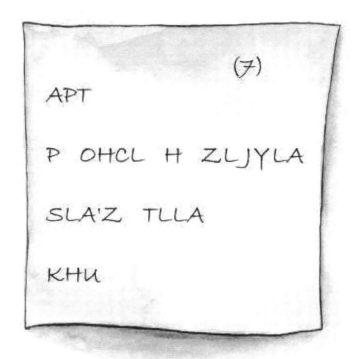

2. Decrypt Dan's second note to Tim.

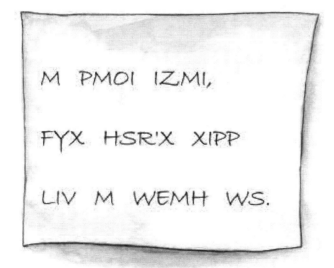

(Text page 22)

3. Decrypt each answer by first figuring out the keys. Let the one-letter words help you.

   a. **Riddle:** What do you call a happy Lassie?
   Answer:

E		N	S	P	P	C		G	S	P	P	M	I														

   b. **Riddle:** Knock, knock. *Who's there?* Cash. *Cash who?*
   Answer:

O		Q	T	K	C		E	U	A		C	K	X	K		Y	U	S	K								

Q	O	T	J		U	L		T	A	Z																	

   c. **Riddle:** What's the noisiest dessert?
   Answer:

W		G	Q	F	S	O	A																				

4. Decrypt the following quotation:

H	S		R	S	X		A	S	V	V	C		E	F	S	Y	X		C	S	Y	V					

H	M	J	J	M	G	Y	P	X	M	I	W		M	R													

Q	E	X	L	I	Q	E	X	M	G	W	,		M		E	W	W	Y	V	I							

C	S	Y		X	L	E	X		Q	M	R	I		E	V	I											

K	V	I	E	X	I	V	.																				

   —Albert Einstein

© 2006 A K Peters, Ltd., Wellesley, MA

The Cryptoclub: Using Mathematics to Make and Break Secret Codes

Chapter 3: Breaking Caesar Ciphers

(Text page 24)

Decrypt each of the following quotations. Tell the key used to encrypt.

5.

| P | K | B | | K | X | N | | K | G | K | I | | D | R | O | | L | O | C | D | |

| Z | B | S | J | O | | D | R | K | D | | V | S | P | O | | Y | P | P | O | B | C |

| S | C | | D | R | O | | M | R | K | X | M | O | | D | Y | | G | Y | B | U | |

| R | K | B | N | | K | D | | G | Y | B | U | | G | Y | B | D | R | |

| N | Y | S | X | Q | . | |

—Theodore Roosevelt   Key = _____

6.

| J | A | J | S | | N | K | | D | T | Z | ' | W | J | | T | S | | Y | M | J | |

| W | N | L | M | Y | | Y | W | F | H | P | , | | D | T | Z | ' | Q | Q | | L | J | Y |

| W | Z | S | | T | A | J | W | | N | K | | D | T | Z | | O | Z | X | Y | |

| X | N | Y | | Y | M | J | W | J | . | |

—Will Rogers   Key = _____

© 2006 A K Peters, Ltd., Wellesley, MA

The Cryptoclub: Using Mathematics to Make and Break Secret Codes

(Text page 25)

7.

R	C	A	B		J	M	K	I	C	A	M		A	W	U	M	B	P	Q	V	O

L	W	M	A	V	'	B		L	W		E	P	I	B		G	W	C

X	T	I	V	V	M	L		Q	B		B	W		L	W		L	W	M	A	V	'	B

U	M	I	V		Q	B	'	A		C	A	M	T	M	A	A	.

—Thomas A. Edison   Key = _____

8.

Q	B	A	'	G		J	N	Y	X		O	R	U	V	A	Q		Z	R	,		V

Z	N	L		A	B	G		Y	R	N	Q	.		Q	B	A	'	G		J	N	Y	X

V	A		S	E	B	A	G		B	S		Z	R	,		V		Z	N	L

A	B	G		S	B	Y	Y	B	J	.		W	H	F	G		J	N	Y	X

O	R	F	V	Q	R		Z	R		N	A	Q		O	R		Z	L

S	E	V	R	A	Q	.

—Albert Camus   Key = _____

© 2006 A K Peters, Ltd., Wellesley, MA

The Cryptoclub: Using Mathematics to Make and Break Secret Codes

Chapter 3: Breaking Caesar Ciphers

Name _____ Date _____

(Text page 25)

9.

| O | C | P | A | Q | H | N | K | H | G' | U | H | C | K | N | W | T | G | U | C | T | G | R |

| G | Q | R | N | G | Y | J | Q | F | K | F | P | Q | V | T | G | C | N | K | B | G | J | Q |

| Y | E | N | Q | U | G | V | J | G | A | Y | G | T | G | V | Q | U | W | E | E | G | U | U |

| Y | J | G | P | V | J | G | A | I | C | X | G | W | R | . |

—Thomas A. Edison   Key = _____

10. **Challenge.**

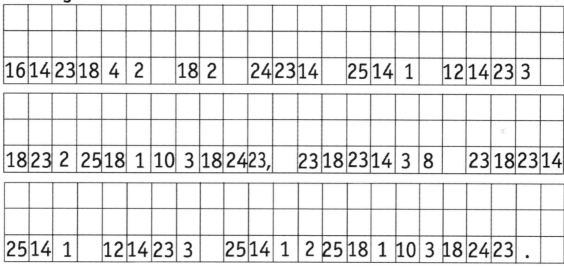

| 16 | 14 | 23 | 18 | 4 | 2 | | 18 | 2 | | 24 | 23 | 14 | | 25 | 14 | 1 | | 12 | 14 | 23 | 3 |

| 18 | 23 | 2 | | 25 | 18 | 1 | 10 | 3 | 18 | 24 | 23, | | 23 | 18 | 23 | 14 | 3 | 8 | | 23 | 18 | 23 | 14 |

| 25 | 14 | 1 | | 12 | 14 | 23 | 3 | | 25 | 14 | 1 | 2 | 25 | 18 | 1 | 10 | 3 | 18 | 24 | 23 | . |

—Thomas A. Edison   Key = _____

© 2006 A K Peters, Ltd., Wellesley, MA

The Cryptoclub: Using Mathematics to Make and Break Secret Codes

You can use this page for your own messages.

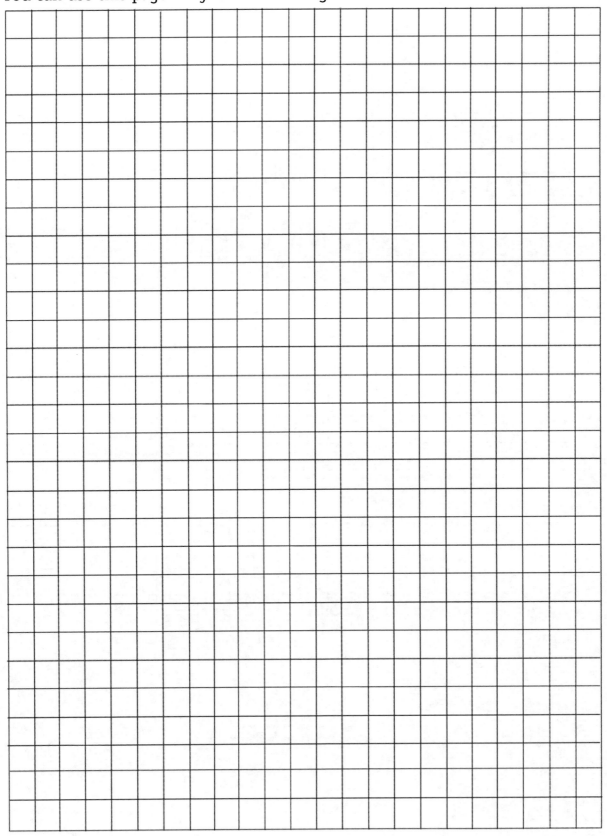

Chapter 3: Breaking Caesar Ciphers

© 2006 A K Peters, Ltd., Wellesley, MA

The Cryptoclub: Using Mathematics to Make and Break Secret Codes

© 2006 A K Peters, Ltd., Wellesley, MA

Name _____ Date_____

Chapter 4: Keyword Ciphers
(Text page 31)

Write the keyword ciphers in the tables. Decrypt the answers to the riddles.

1. Keyword: DAN, Key letter: h

a	b	c	d	e	f	g	h	i	j	k	l	m	n	o	p	q	r	s	t	u	v	w	x	y	z

Riddle: What is worse than biting into an apple and finding a worm?

Answer: YAFWAFZ DTCY T PGJE.

2. Keyword: HOUSE, Key letter: m

a	b	c	d	e	f	g	h	i	j	k	l	m	n	o	p	q	r	s	t	u	v	w	x	y	z

Riddle: Is it hard to spot a leopard?

Answer: OU. CVQJ LAQ MUAO CVLC GLJ.

3. Keyword: MUSIC, Key letter: d

a	b	c	d	e	f	g	h	i	j	k	l	m	n	o	p	q	r	s	t	u	v	w	x	y	z

Riddle: What part of your body has the most rhythm?

Answer: VHPL UXLMLPFN

The Cryptoclub: Using Mathematics to Make and Break Secret Codes

(Text page 31)

4.   Keyword: FISH, Key letter: a

a	b	c	d	e	f	g	h	i	j	k	l	m	n	o	p	q	r	s	t	u	v	w	x	y	z

Riddle: What does Mother Earth use for fishing?

Answer: TDA MNQTD FMH RNUTD ONKAR

5.   Keyword: ANIMAL, Key letter: g

a	b	c	d	e	f	g	h	i	j	k	l	m	n	o	p	q	r	s	t	u	v	w	x	y	z

Riddle: Why was the belt arrested?

Answer: ZEH NEBXIDA OF KNY FUDKJ.

6.   Keyword: RABBIT, Key letter: f

a	b	c	d	e	f	g	h	i	j	k	l	m	n	o	p	q	r	s	t	u	v	w	x	y	z

Riddle: How do rabbits travel?

Answer: WS BVKZHDVFZ

7.   Keyword: MISSISSIPPI, Key letter: d

a	b	c	d	e	f	g	h	i	j	k	l	m	n	o	p	q	r	s	t	u	v	w	x	y	z

Riddle: What ears cannot hear?

Answer: IXLN HS ZHLG

© 2006 A K Peters, Ltd., Wellesley, MA

The Cryptoclub: Using Mathematics to Make and Break Secret Codes

© 2006 A K Peters, Ltd., Wellesley, MA

The Cryptoclub: Using Mathematics to Make and Break Secret Codes

Name _____ Date_____

(Text page 32)

8. Keyword: SKITRIP, Key letter: p
   (It is a long message, so you may want to share the work with a group.)

a	b	c	d	e	f	g	h	i	j	k	l	m	n	o	p	q	r	s	t	u	v	w	x	y	z

[1] OLIL FIL ROL JLRFQWT ZM ROL ZPRJZZI

[2] HWPG'T TVQ RIQS: ROL RBZ-JFD RIQS RZ

[3] SQYL XZPYRFQY BQWW GL TFRPIJFD FYJ

[4] TPYJFD, ROL MQITR BLLVLYJ QY MLGIPFID.

[5] ROL GPT BQWW WLFAL MIZX ROL SFIV'T

[6] OLFJKPFIRLIT FR LQNOR FX FYJ ILRPIY FR

[7] RLY SX TPYJFD.

[8] ILNQTRIFRQZY MZIXT FIL JPL GD YLCR MIQJFD

[9] SQHV ROLX PS QY ROL SFIV ZMMQHL.

[10] ROL RIQS QT WQXQRLJ RZ ROL MQITR

[11] RBLYRD BOZ TQNY PS, TZ SWLFTL OPIID ZI

[12] ROLIL XQNOR YZR GL LYZPNO TSFHL.

(Text page 32)

9. Create your own keyword cipher.

Keyword: _____   Key letter: _____

a	b	c	d	e	f	g	h	i	j	k	l	m	n	o	p	q	r	s	t	u	v	w	x	y	z

On your own paper, encrypt a message to another group. Tell them your keyword and key letter so they can decrypt.

Here are extra tables to use when encrypting and decrypting other messages.

a	b	c	d	e	f	g	h	i	j	k	l	m	n	o	p	q	r	s	t	u	v	w	x	y	z

a	b	c	d	e	f	g	h	i	j	k	l	m	n	o	p	q	r	s	t	u	v	w	x	y	z

a	b	c	d	e	f	g	h	i	j	k	l	m	n	o	p	q	r	s	t	u	v	w	x	y	z

a	b	c	d	e	f	g	h	i	j	k	l	m	n	o	p	q	r	s	t	u	v	w	x	y	z

a	b	c	d	e	f	g	h	i	j	k	l	m	n	o	p	q	r	s	t	u	v	w	x	y	z

© 2006 A K Peters, Ltd., Wellesley, MA

© 2006 A K Peters, Ltd., Wellesley, MA

The Cryptoclub: Using Mathematics to Make and Break Secret Codes

Name _____ Date _____

Chapter 5: Letter Frequencies
(Text page 37)

CLASS ACTIVITY: Finding Relative Frequencies of Letters in English

Part 1. Collecting data from a small sample.

    a. Choose about 100 English letters from a newspaper or other English text. (Note: If you are working without a class, choose a larger sample—around 500 letters. Then skip Parts 1 and 2.)

    b. Work with your group to count the **A**s, **B**s, etc., in your sample.

    c. Enter your data in the table below.

Letter Frequencies for Your Sample

Letter	Frequency	Letter	Frequency
A		N	
B		O	
C		P	
D		Q	
E		R	
F		S	
G		T	
H		U	
I		V	
J		W	
K		X	
L		Y	
M		Z	

Part 2. Combining data to make a larger sample.

    a. Record your data from Part 1 on your class's Class Letter Frequencies table. (Your teacher will provide this table on the board, overhead, or chart paper.)

    b. Your teacher will assign your group a few rows to add. Enter your sums in the group table.

Name _____ Date _____

(Text pages 37–38)

Part 3. Computing relative frequencies.

Enter your class's combined data from the "Total for All Groups" column of Part 2 into the "Frequency" column. Then compute the relative frequencies.

Letter	Frequency	Relative Frequency		
		Fraction	Decimal (to 3 places)	Percent (%) (to nearest tenth)
A				
B				
C				
D				
E				
F				
G				
H				
I				
J				
K				
L				
M				
N				
O				
P				
Q				
R				
S				
T				
U				
V				
W				
X				
Y				
Z				
Total				

© 2006 A K Peters, Ltd., Wellesley, MA

The Cryptoclub: Using Mathematics to Make and Break Secret Codes

Name _____ **Date** _____

1. a. What percent of the letters in the class sample were the letter **T**? ____%

    b. About how many **T**s would you expect in a sample of 100 letters? _____

    c. If your sample was about 100 letters, was your answer to 1b close to the number of **T**s you found in your sample? _____

2. a. What percent of the letters in the class sample were the letter **E**? ____%

    b. About how many **E**s would you expect in a sample size of 100? _____

    c. About how many **E**s would you expect in a sample of 1000 letters? ____

3. Arrange the letters in your class table in order, from most common to least common.

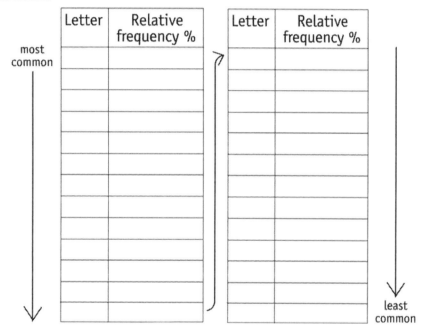

4. The table on Page 39 of the text shows frequencies of letters in English computed using a sample of about 100,000 letters. How is your class data the same as the data in that table? How is it different? Why might it be different?

    _____

    _____

    _____

    _____

© 2006 A K Peters, Ltd., Wellesley, MA

The Cryptoclub: Using Mathematics to Make and Break Secret Codes

Chapter 6: Breaking Substitution Ciphers
(Text page 49)

1. Use frequency analysis to decrypt Jenny's message, which is shown on the following page.

   a. Record the number of occurrences (frequency) of each letter in her message. Then compute the relative frequencies.

Letter Frequencies for Jenny's Message

Letter	Frequency	Relative Frequency		
		Fraction	Decimal (to 3 places)	Percent (%) (to nearest tenth)
A				
B				
C				
D				
E				
F				
G				
H				
I				
J				
K				
L				
M				
N				
O				
P				
Q				
R				
S				
T				
U				
V				
W				
X				
Y				
Z				
Total				

© 2006 A K Peters, Ltd., Wellesley, MA

The Cryptoclub: Using Mathematics to Make and Break Secret Codes

Name _____ Date _____

b. Arrange letters in order from the most common to the least common.

In Message		In English	
Letter	Rel. Freq. (%)	Letter	Rel. Freq. (%)
		e	12.7
		t	9.1
		a	8.2
		o	7.5
		i	7.0
		n	6.7
		s	6.3
		h	6.1
		r	6.0
		d	4.3
		l	4.0
		c	2.8
		u	2.8
		m	2.4
		w	2.4
		f	2.2
		g	2.0
		y	2.0
		p	1.9
		b	1.5
		v	1.0
		k	0.8
		j	0.2
		q	0.1
		x	0.1
		z	0.1

c. Now decrypt Jenny's message, using the frequencies to help you guess the correct substitutions. Record your substitutions in the Substitution Table below the message. Tip: Use pencil!

Jenny's Message

[1] Y XTNDQ DNQYS EFNFYSU

[2] JKLM NUUSGUPTQ FXTL JYII

[3] WYHT NJNL VDTT PYDPGE

[4] FYPCTFE FS FXT VYDEF

[5] FJTUFL-VYHT ATSAIT JXS

[6] PNII YU. YF ESGUQE IYCT

[7] VGU. ITF'E NII PNII NUQ

[8] WS FSWTFXTD.

Substitution Table

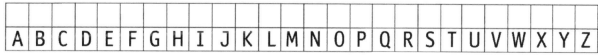

A	B	C	D	E	F	G	H	I	J	K	L	M	N	O	P	Q	R	S	T	U	V	W	X	Y	Z

© 2006 A K Peters, Ltd., Wellesley, MA

The Cryptoclub: Using Mathematics to Make and Break Secret Codes

Name _____ Date _____

(Text page 49)

2. Here is another message to decrypt using frequency analysis. The relative frequencies have been computed for you. Record your substitutions in the Substitution Table below the message. Tip: Use pencil!

In Message 2

Letter	Rel. Freq. (%)
D	11.4
G	9.8
Q	8.3
T	7.8
C	6.7
K	6.7
E	6.2
L	5.7
N	5.7
S	5.2
I	3.6
U	3.6
J	3.1
M	2.6
Y	2.6
A	1.6
B	1.6
H	1.6
R	1.6
W	1.6
O	1.0
Z	1.0
V	0.5
X	0.5
F	0.0
P	0.0

In English

Letter	Rel. Freq. (%)
e	12.7
t	9.1
a	8.2
o	7.5
i	7.0
n	6.7
s	6.3
h	6.1
r	6.0
d	4.3
l	4.0
c	2.8
u	2.8
m	2.4
w	2.4
f	2.2
g	2.0
y	2.0
p	1.9
b	1.5
v	1.0
k	0.8
j	0.2
q	0.1
x	0.1
z	0.1

Message 2

[1] BQGKNJG SDKT CDQ MGVLQETD

[2] BQGKNSLK G CGKNSLJD KDW

[3] SCEQT MLQ CES REQTCNGY.

[4] UKMLQTUKGTDIY, ET CGN G

[5] SEZD MLUQTDDK ALIIGQ GKN

[6] TCD RLY CGN G SEZD SEXTDDK

[7] KDAH. CD NUTEMUIIY WQLTD CDQ,

[8] "NDGQ BQGJJY, TCGKHS CDGOS.

[9] E'N WQETD JLQD RUT E'J GII

[10] ACLHDN UO."

Substitution Table

A	B	C	D	E	F	G	H	I	J	K	L	M	N	O	P	Q	R	S	T	U	V	W	X	Y	Z

© 2006 A K Peters, Ltd., Wellesley, MA

The Cryptoclub: Using Mathematics to Make and Break Secret Codes

Chapter 6: Breaking Substitution Ciphers

© 2006 A K Peters, Ltd., Wellesley, MA

The Cryptoclub: Using Mathematics to Make and Break Secret Codes

Name _____ Date _____

Chapter 7: Combining Caesar Ciphers
(Text pages 56–57)

1. Encrypt using a Vigenère cipher with keyword DOG.

keyword:
plaintext: h i d d e n   t r e a s u r e
ciphertext:

2. Encrypt using a Vigenère cipher with keyword CAT.

M e e t   m e   t o n i g h t   a t

m i d n i g h t

Return to Text

3. Decrypt using a Vigenère cipher with keyword CAT.

Q K,   U W T   P J E K G   S A C L E   Y E   F G E M?

4. Decrypt using a Vigenère cipher with keyword LIE.

L   T M P   K E Y   B V L D I W   P E W N A L G

E C W Y Y L   X S M   A Z Z P O   E L T T I   E P I

E Z Y E P   M D   X Y E B M Y O   S Y   Q X D

A L Z M W.

—Mark Twain

(Text page 58)

5. Use the Vigenère square—not a cipher wheel—to finish encrypting:

D	O	G		D	O	G	D	O	G		D	O	G	D	O	G	D	O	G	D	O				
t	o	p		s	e	c	r	e	t		i	n	f	o	r	m	a	t	i	o	n				
W	C																								

6. Use the Vigenère square to decrypt the following. (Keyword: BLUE)

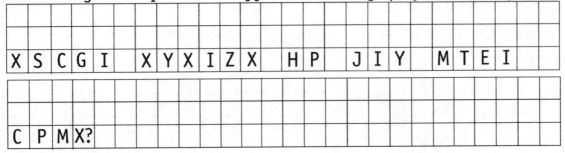

| X | S | C | G | I | | X | Y | X | I | Z | X | | H | P | | J | I | Y | | M | T | E | I |

| C | P | M | X? |

7. Use either the cipher-wheel method or the Vigenère-square method to decrypt the following quotations from author Mark Twain.

a. Keyword: SELF

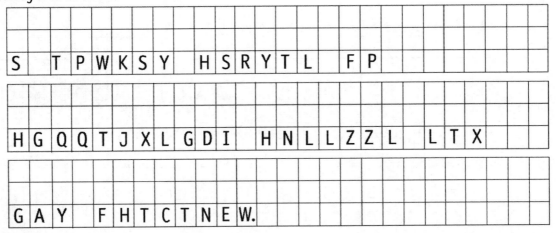

| S | T | P | W | K | S | Y | | H | S | R | Y | T | L | | F | P |

| H | G | Q | Q | T | J | X | L | G | D | I | | H | N | L | L | Z | Z | L | | L | T | X |

| G | A | Y | | F | H | T | C | T | N | E | W. |

b. Keyword: READ

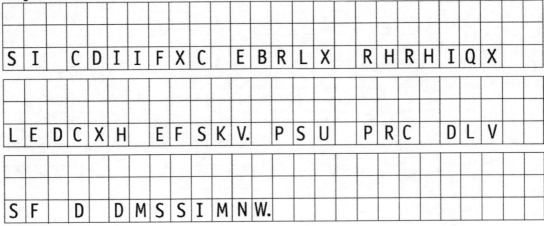

| S | I | | C | D | I | I | F | X | C | | E | B | R | L | X | | R | H | R | H | I | Q | X |

| L | E | D | C | X | H | | E | F | S | K | V. | | P | S | U | | P | R | C | | D | L | V |

| S | F | | D | | D | M | S | S | I | M | N | W. |

© 2006 A K Peters, Ltd., Wellesley, MA

The Cryptoclub: Using Mathematics to Make and Break Secret Codes

© 2006 A K Peters, Ltd., Wellesley, MA

Name _____   **Date** _____

(Text page 59)

8. Use either the cipher-wheel method or the Vigenère-square method to decrypt the following quotes from Mark Twain.

a. Keyword: CAR

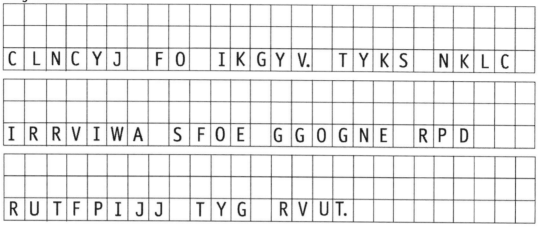

b. Keyword: TWAIN

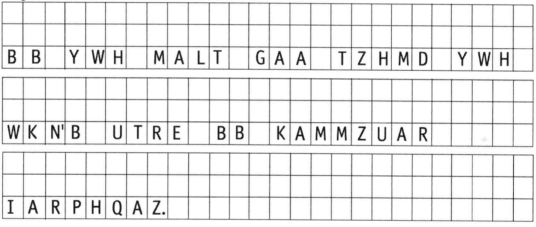

c. Keyword: NOT

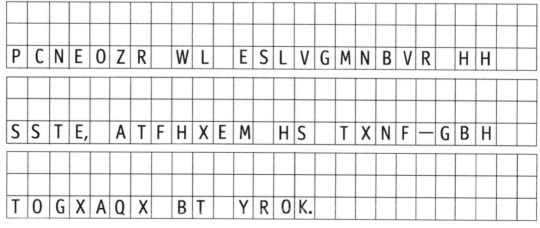

(Text page 60)

9. Use either the cipher-wheel method or the Vigenère-square method to decrypt the following quotations.

a. Keyword: WISE

D W F I O B Q   M O   B Z I   B Q J W P

K Z E L B W V   E V   L L A   J G S G   W X

A E A V S I.

—Thomas Jefferson

b. Keyword: STONE

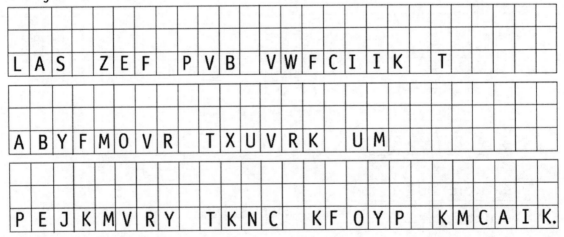

L A S   Z E F   P V B   V W F C I I K   T

A B Y F M O V R   T X U V R K   U M

P E J K M V R Y   T K N C   K F O Y P   K M C A I K.

—Chinese Proverb

© 2006 A K Peters, Ltd., Wellesley, MA

The Cryptoclub: Using Mathematics to Make and Break Secret Codes

© 2006 A K Peters, Ltd., Wellesley, MA

The Cryptoclub: Using Mathematics to Make and Break Secret Codes

Name_____ **Date**_____

(Text page 60)

10. Find a quote from a famous person. Encrypt it using a Vigenère Cipher. Use it to play Cipher Tag.

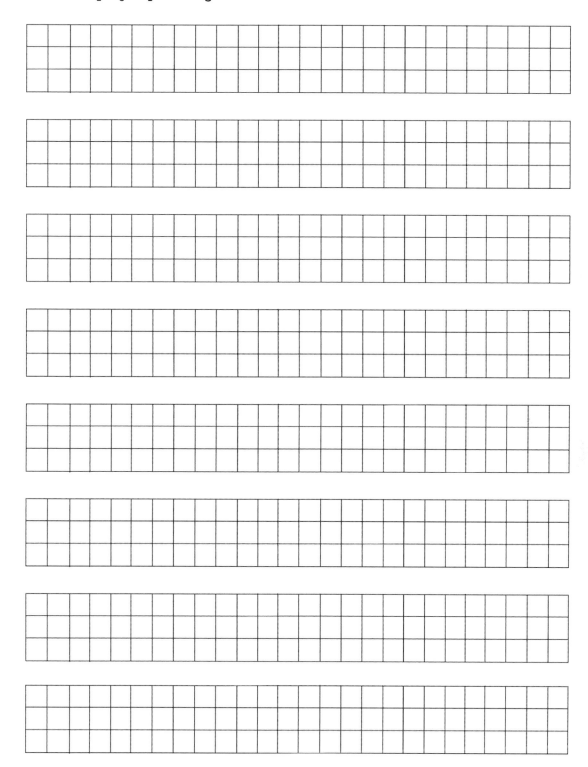

(Text page 60)

11. **Challenge.** Explore how to describe a Vigenère cipher using numbers. In Chapter 2, you worked with number messages. You described Caesar ciphers with arithmetic—by adding to encrypt and subtracting to decrypt. The Vigenère Cipher can be described with arithmetic too. Instead of writing the keyword repeatedly, change the letters of the keyword to numbers and write the numbers repeatedly. Then add to encrypt. For an example, see page 60 of the text. Encrypt and decrypt your own message with this method.

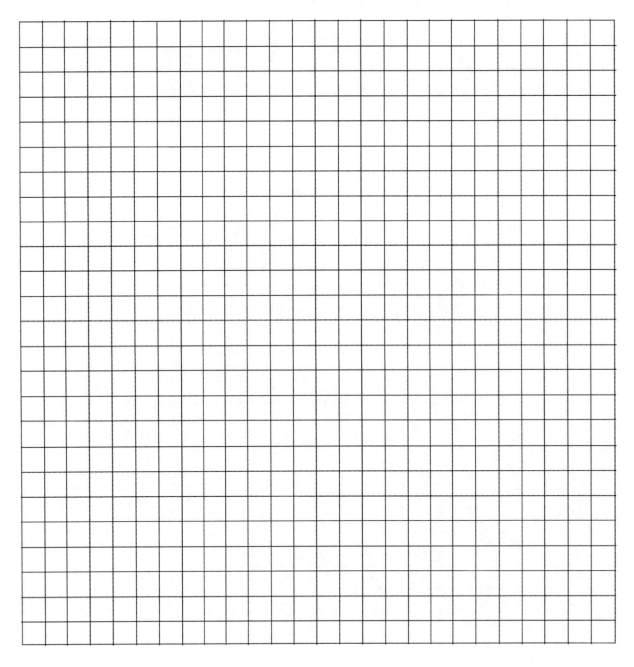

© 2006 A K Peters, Ltd., Wellesley, MA

The Cryptoclub: Using Mathematics to Make and Break Secret Codes

Chapter 8: Cracking Vigenère Ciphers When You Know the Key Length
(Text pages 72–73)

CLASS ACTIVITY. Finish Decrypting the Girls' Message

Finish decrypting the Girls' Message (key length 4) on pages W36–W37. Your teacher will assign your group 3 or 4 lines of the message to work with.

1. **First wheel.** The letters for the first wheel are already decrypted. What letter was matched with **a**? _____

2. **Second wheel**
   a. Use the table on page 72 of the text to decide how to turn the second wheel. Then decrypt the letters with 2 underneath in your assigned lines.
   b. What letter did you match with **a**? _____

3. **Third wheel**
   a. Find the number of **A**s, **B**s, **C**s, etc. among the letters with 3 underneath. Record your data in the tables on page W38.
   b. Use the class data from 3a to decide how to turn the third wheel. Then decrypt the letters with 3 underneath in your assigned lines.
   c. What letter did you match with **a**? _____

4. **Fourth wheel**
   a. Use the partly decrypted message to guess how to decrypt one of the letters with 4 underneath. Use this to figure out what the fourth wheel must be. Then decrypt the rest of your assigned lines.
   b. What letter did you match with **a**? _____

5. What was the keyword? ____ ____ ____ ____

© 2006 A K Peters, Ltd., Wellesley, MA

The Cryptoclub: Using Mathematics to Make and Break Secret Codes

(Text page 70–73)

The Girls' Message

[1]
```
t      e       a       t        o        m
W P Q V H   E M W   D   T U X W T Q   F R G   Z E P M P
1 2 3 4 1   2 3 4   1   2 3 4 1 2 3   4 1 2   3 4 1 2 3
```

[2]
```
   e      .      t        i         y        t
N H A E I.   W P Q   F L O   N S B A   U R   W P Q
4 1 2 3 4    1 2 3   4 1 2   3 4 1 2   3 4   1 2 3
```

[3]
```
   e       b       o          o         a         y          s
R H Q S L E W D L R W P   G R V E X D V F P B   B Q E V M P
4 1 2 3 4 1 2 3 4 1 2 3   4 1 2 3 4 1 2 3 4 1   2 3 4 1 2 3
```

[4]
```
   i  .      m       m          h         f        e          m
L L U.   E S P M F M P M E   X K M K   S I N Q V H L   T M P
4 1 2    3 4 1 2 3 4 1 2 3   4 1 2 3   4 1 2 3 4 1 2   3 4 1
```

[5]
```
      o       b       e          n         e          d
I   O L R Q O I   E M F A H M Z   E   Q Q O O H T   M R G
2   3 4 1 2 3 4   1 2 3 4 1 2 3   4   1 2 3 4 1 2   3 4 1
```

[6]
```
      m  .      s       l          s          k
I   P M P M.   V I V A Q   E O E M C V   B A S N   B T I
2   3 4 1 2    3 4 1 2 3   4 1 2 3 4 1   2 3 4 1   2 3 4
```

[7]
```
n         e     -       t          l  ,           w          i         r.
Q Q O O H T   - M J W M D   E O T,   U X   Z I E   F L O S I U.
1 2 3 4 1 2   3 4 1 2 3     4 1 2    3 4   1 2 3   4 1 2 3 4 1
```

[8]
```
      b       o       a          e         d
B T I   E Q S   F R G E   P D C S L H L   M R G
2 3 4   1 2 3   4 1 2 3   4 1 2 3 4 1 2   3 4 1
```

[9]
```
      g      .      o       a        t          e
T M Y J P Q H.   R V Q   H D G   M J W M D   N H A E I
2 3 4 1 2 3 4    1 2 3   4 1 2   3 4 1 2 3   4 1 2 3 4
```

[10]
```
g         b       h       c         ,         h         a         r
J Z M F E M P   X K M   Z M F S Q P,   K Q E   J D B T I U
1 2 3 4 1 2 3   4 1 2   3 4 1 2 3 4    1 2 3   4 1 2 3 4 1
```

(continued)

© 2006 A K Peters, Ltd., Wellesley, MA

The Cryptoclub: Using Mathematics to Make and Break Secret Codes

Chapter 8: Cracking Vigenère Ciphers When You Know the Key length

Name _____ Date_____

(Text page 70-73)

[11]
- decoded: k (N) … a … e … s … "j … e,
- cipher: B A S N   P U Q   D A U H H   I Z H   V I U H,   "M M E W H,
- key: 2 3 4 1   2 3 4   1 2 3 4 1   2 3 4   1 2 3 4   1 2 3 4 1

[12]
- decoded: s … y … e … i … u
- cipher: B T S V M   N S B A   M V H   U M O L V S   J X V   A J
- key: 2 3 4 1 2   3 4 1 2   3 4 1   2 3 4 1 2 3   4 1 2   3 4

[13]
- decoded: y … . … h … h … y … o … ' … n
- cipher: B W G.   X K M K   X K Q Z O   B W G   H R V' F   O Q W I
- key: 1 2 3   4 1 2 3   4 1 2 3 4   1 2 3   4 1 2 3   4 1 2 3

[14]
- decoded: h … m … w … h … e … n
- cipher: X K M   P M P M   U W   Z W D X K   U A V H   B T E Q
- key: 4 1 2   3 4 1 2   3 4   1 2 3 4 1   2 3 4   2 3 4 1

[15]
- decoded: n … e ." … s … r … e … d
- cipher: B T I   Q Q O O H T."   V I V A Q   K U Q Z R H L   M R G
- key: 2 3 4   1 2 3 4 1 2   3 4 1 2 3   4 1 2 3 4 1 2   3 4 1

[16]
- decoded: d, … " … ' t … r … d. … o
- cipher: A M M G,   "L A R' W   E A V U G   P E G.   Q   W R R E
- key: 2 3 4 1   2 3 4 1   2 3 4 1 2   3 4 1   2   3 4 1 2

[17]
- decoded: i … s … t … r … . … t
- cipher: I L L K T   M V   E A V W P   Y S U M.   N Y W   Q R   M
- key: 3 4 1 2 3   4 1   2 3 4 1 2   3 4 1 2   3 4 1   2 3   4

[18]
- decoded: t … t … i … , … h … o … s
- cipher: W W A O   W P Q   H L U Q,   X K M K   A R C X H   V B A T
- key: 1 2 3 4   1 2 3   4 1 2 3   4 1 2 3   4 1 2 3 4   1 2 3 4

[19]
- decoded: d … g … . … o … i ' … o … c
- cipher: G W U R J   Q F.   W R   N M V   L' D Q   G R T X I F   B Q H
- key: 1 2 3 4   2 3   4 1   2 3 4   1 2 3   4 1 2 3 4   1 2 3 4

[20]
- decoded: t … o … r … ."
- cipher: W M Z   H R T X E U A."
- key: 1 2 3   4 1 2 3 4 1 2

© 2006 A K Peters, Ltd., Wellesley, MA

The Cryptoclub: Using Mathematics to Make and Break Secret Codes

Name _____ **Date** _____

(Text page 70–73)

Tables for the third wheel of the Girls' Message.

What line numbers are assigned to your group? _____

To save work, count the letters in your assigned lines only. Then combine data with your class to get a total.

Frequency in Your Assigned Lines				Class Total	
Wheel 3 letter	Tally (optional)	Number in your lines only		Wheel 3 letter	Number in entire message
A				A	
B				B	
C				C	
D				D	
E				E	
F				F	
G				G	
H				H	
I				I	
J				J	
K				K	
L				L	
M				M	
N				N	
O				O	
P				P	
Q				Q	
R				R	
S				S	
T				T	
U				U	
V				V	
W				W	
X				X	
Y				Y	
Z				Z	

© 2006 A K Peters, Ltd., Wellesley, MA

The Cryptoclub: Using Mathematics to Make and Break Secret Codes

Chapter 8: Cracking Vigenère Ciphers When You Know the Key length

© 2006 A K Peters, Ltd., Wellesley, MA

Name _____ **Date** _____

Chapter 9: Factoring
(Text page 76)

1. Find all factors of the following numbers:

   a. 15 _____

   b. 24 _____

   c. 36 _____

   d. 60 _____

   e. 23 _____

2. List four multiples of 5.

   _____

3. List all prime numbers less than 30.

   _____

4. List all composite numbers from 30 to 40.

   _____

The Cryptoclub: Using Mathematics to Make and Break Secret Codes

© 2006 A K Peters, Ltd., Wellesley, MA

Name _____ **Date** _____

(Text page 77)

5. Use a factor tree to find the prime factorization of each of the following
   numbers:

a.

24

a. 24 = _____

b.

56

b. 56 = _____

c.

90

c. 90 = _____

The Cryptoclub: Using Mathematics to Make and Break Secret Codes

© 2006 A K Peters, Ltd., Wellesley, MA

Name _____ **Date** _____

(Text page 79)

6. Circle the numbers that are divisible by 2. How do you know?

    a. 284 _____

    b. 181 _____

    c. 70 _____

    d. 5456 _____

7. Circle the numbers that are divisible by 3. How do you know?

    a. 585 _____

    b. 181 _____

    c. 70 _____

    d. 6249 _____

8. Circle the numbers that are divisible by 4. How do you know?

    a. 348 _____

    b. 236 _____

    c. 621 _____

    d. 8480 _____

9. Circle the numbers that are divisible by 5. How do you know?

    a. 80 _____

    b. 995 _____

    c. 232 _____

    d. 444 _____

The Cryptoclub: Using Mathematics to Make and Break Secret Codes

(Text page 79)

10. Circle the numbers that are divisible by 6. How do you know?

a. 96 _____

b. 367 _____

c. 642 _____

d. 842 _____

11. Circle the numbers that are divisible by 9. How do you know?

a. 333 _____

b. 108 _____

c. 348 _____

d. 1125 _____

12. Circle the numbers that are divisible by 10. How do you know?

a. 240 _____

b. 1005 _____

c. 60 _____

d. 9900 _____

© 2006 A K Peters, Ltd., Wellesley, MA

The Cryptoclub: Using Mathematics to Make and Break Secret Codes

Name _____ Date _____

(Text page 81)

13. Use a factor tree to find the prime factorization of each of the following numbers. Write each factorization using exponents.

    a.

$$2430$$

a. 2430 = _____

    b.

$$4680$$

b. 4680 = _____

© 2006 A K Peters, Ltd., Wellesley, MA

The Cryptoclub: Using Mathematics to Make and Break Secret Codes

(Text page 81)

13. c.

c. 357 = _____

d.

56,133

d. 56,133 = _____

© 2006 A K Peters, Ltd., Wellesley, MA

The Cryptoclub: Using Mathematics to Make and Break Secret Codes

© 2006 A K Peters, Ltd., Wellesley, MA

The Cryptoclub: Using Mathematics to Make and Break Secret Codes

Name _____ **Date** _____

(Text page 81)

13. e.

14,625

e. 14,625 = _____

f.

8550

f. 8550 = _____

(Text page 82)

14. Find the common factors of the following pairs of numbers:

     a. 10 and 25   Common factors: _____

     b. 12 and 18   Common factors: _____

     c. 45 and 60   Common factors: _____

15. Find the greatest common factor of each of the following pairs of numbers:

     a. 12 and 20   Greatest common factor: _____

     b. 50 and 75   Greatest common factor: _____

     c. 30 and 45   Greatest common factor: _____

© 2006 A K Peters, Ltd., Wellesley, MA

© 2006 A K Peters, Ltd., Wellesley, MA

Name _____ **Date** _____

(Text page 82)

16. For each list of numbers, factor the numbers into primes and then find all common factors for the list. Use the space beside the problems for any factor trees you want to make.

    a. 14 = _____

         22 = _____

         10 = _____

         Common factor(s): _____

    b. 66 = _____

         210 = _____

         180 = _____

         Common factor(s): _____

    c. 30 = _____

         90 = _____

         210 = _____

         Common factor(s): _____

Continue to the next chapter.

© 2006 A K Peters, Ltd., Wellesley, MA

The Cryptoclub: Using Mathematics to Make and Break Secret Codes

© 2006 A K Peters, Ltd., Wellesley, MA

Name _____ Date_____

Chapter 10: Using Common Factors to Crack Vigenère Ciphers
(Text page 88)

These problems involve entries Meriwether Lewis wrote in his journal during the Lewis and Clark Expedition. (You might notice that the spelling is not always the same as modern-day spelling, but we show it as it originally was written.)

1. *Sunday, May 20, 1804*

   "We set forward... to join my friend companion and fellow labourer Capt. William Clark, who had previously arrived at that place with the party destined for the discovery of the interior of the continent of North America.... As I had determined to reach St. Charles this evening and knowing that there was now no time to be lost I set forward in the rain... and joined Capt Clark, found the party in good health and sperits."

   a. Circle all occurrences of **the** in the message above. Include examples such as "**the**re" in which **the** occurs as part of a word.

   b. Find the distance from the beginning of the last **the** in the 5th line to the beginning of **the** in the 6th line. _____

   c. Choose a keyword from RED, BLUE, ARTICHOKES, TOMATOES that will encrypt in exactly in the same way the two occurrences of **the** in the following phrase (from the last sentence of the message). Then, use it to encrypt the phrase.

   > **Note:**
   > To find the distance between repeated strings of letters, count the letters from the beginning of the first string up to (but not including) the beginning of the second. (Don't count punctuation or spaces.)
   >
   > For example, in **XYZABCDXYZ**, the strings **XYZ** are a distance of 7 letters apart as counted here:
   >
   > **XYZABCDXYZ**
   > **1234567**

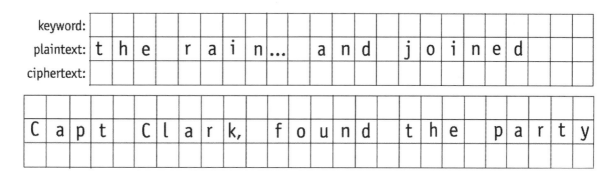

The Cryptoclub: Using Mathematics to Make and Break Secret Codes

(Text page 88)

1. d. Choose a keyword from RED, BLUE, ARTICHOKES, TOMATOES that will encrypt in different ways the two occurrences of **the** in the phrase. Then use it to encrypt.

keyword:																			
plaintext:	t	h	e		r	a	i	n...		a	n	d		j	o	i	n	e	d
ciphertext:																			

C	a	p	t		C	l	a	r	k,		f	o	u	n	d		t	h	e		p	a	r	t	y

e. Of the keywords you have not used, which would encrypt the two occurrences of **the** in the phrase above

in the same way? _____

in different ways? _____

Give reasons for your answers:

_____

_____

_____

_____

_____

_____

© 2006 A K Peters, Ltd., Wellesley, MA

The Cryptoclub: Using Mathematics to Make and Break Secret Codes

(Text page 89)

2. *Wednesday, April 7, 1805*

"We were now about to penentrate a country at least two thousand miles [3,219 kilometers] in width, on which the foot of civilized man had never trodden; the good or evil it had in store for us was for experiment yet to determine, and these little vessells contained every article by which we were to expect to subsist or defend ourselves.... I could but esteem this moment of my departure as among the most happy of my life."

a. Circle the occurrences of **the** in the above message.

b. Find the distance from the beginning of **the** in the second line to the beginning of **the** in the third line. _____ List all keyword lengths that would cause these words to be encrypted the same way.

_____

c. Find the distance from the beginning of **the** in the third line to the beginning of **the**se in the fourth line. _____ List all keyword lengths that would cause **the** in these strings to be encrypted the same way.

_____

d. What keyword length(s) would cause all three occurrences of **the** described in 2b and 2c to be encrypted the same way?

_____

© 2006 A K Peters, Ltd., Wellesley, MA

(Text page 89)

2. e. Choose the keyword from the following list that will cause all three occurrences of **the** described in 2b and 2c to be encrypted the same way.

PEAR, APPLE, CARROT, LETTUCE, CUCUMBER, ASPARAGUS, WATERMELON, CAULIFLOWER

f. Write your chosen keyword above the message below. Encrypt each occurrence of **the**. (You don't have to encrypt the entire message.)

keyword:

plaintext: t h e   f o o t   o f   c i v i l i z e d

ciphertext:

m a n   h a d   n e v e r   t r o d d e n;   t h e

g o o d   o r   e v i l   i t   h a d   i n

s t o r e   f o r   u s   w a s   f o r

e x p e r i m e n t   y e t   t o

d e t e r m i n e,   a n d   t h e s e

© 2006 A K Peters, Ltd., Wellesley, MA

The Cryptoclub: Using Mathematics to Make and Break Secret Codes

Continue to the next page for Problem 3.

© 2006 A K Peters, Ltd., Wellesley, MA

(Text page 90)

3. a. Underline strings of letters that repeat in the girls' message below:

[1] WPQVH EMW D TUXWTQ FRG ZEPMP NHAEI. WPQ

[2] FLO NSBA UR WPQ RHQSLEWDLRWP GRVEXDVFPB

[3] BQEVMP LLU. ESPMFMPME XKMK SINQVHL TMP I

[4] OLRQOI EMFAHMZ E QQOOHT MRG I PMPM. VIVAQ

[5] EOEMCV BASN BTI QQOOHT-MJWMD EOT, UX ZIE

[6] FLOSIU. BTI EQS FRGE PDCSLHL MRG TMYJPQH.

[7] RVQ HDG MJWMD NHAEI JZMFEMP XKM ZMFSQP,

[8] KQE JDBTIU BASN PUQ DAUHH IZH VIUH, "MMEWH,

[9] BTSVM NSBA MVH UMOLVS JXV AJ BWG. XKMK

[10] XKQZO BWG HRV'F OQWI XKM PMPM UW ZWDXK

[11] UAVH BTEQ BTI QQOOHT." VIVAQ KUQZRHL MRG

[12] AMMG, "LAR'W EAVUG PEG. Q WRRE ILLKT MV

[13] EAVWP YSUM. NYW QR M WWAO WPQ HLUQ, XKMK

[14] ARCXH VBAT GWURJ QF. WR NMV L'DQ GRTXIFBQH

[15] WMZ HRTXEUA."

Chapter 10: Using Common Factors to Crack Vigenère Ciphers

© 2006 A K Peters, Ltd., Wellesley, MA

The Cryptoclub: Using Mathematics to Make and Break Secret Codes

© 2006 A K Peters, Ltd., Wellesley, MA

The Cryptoclub: Using Mathematics to Make and Break Secret Codes

Name _____ Date _____

(Text page 90)

3. b. Complete the table below. Include the strings shown in the table as well as the strings you found in 3a.

Repeated Strings in the Girls' Message			
Keyword = __DIME__ Key Length = 4			
String	Distance between repetitions	Is key length a factor of distance?	Number of times keyword fits between repetitions
XKM	136	yes	34
XKM	68		
XKM	20		
XKM	100		
ZMF (in line 7)			

c. Is the key length *always, usually,* or *sometimes* a factor of the distance between strings? _____

(Text page 92)

4. a. Underline or circle strings that repeat in Grandfather's message below. Include at least two strings whose distances aren't in the table in 4b. Repetitions of GZS are already underlined.

[1] A VNNS SGIAV GVDJRJ! WG OOF AB

[2] GZS UAZYK PRZWAV HUW HESRVFU

[3] CGGG GB GZS AADVYCA JWIWF NL

[4] HUW BBJHUWFA LWC GT YSYR

[5] KICWFVGF. JZWYW VVCWAY, W SGIAV

[6] GBES FZWAQ GGGBRK. ZNLSE A

[7] PEGITZH GZSZ LC N ESGSZ RPDRJH

[8] GG VNNS GZSZ SDCJOVKSQ. KIEW

[9] SAGITZ, HUWM NJS FAZIWF-VF O

[10] IWFL HIEW TBJA. GZSEW AHKH OW

[11] ABJS-V OWYD FRLIEF OAV GGSYR S

[12] QYSWZ.

Chapter 10: Using Common Factors to Crack Vigenère Ciphers

© 2006 A K Peters, Ltd., Wellesley, MA

The Cryptoclub: Using Mathematics to Make and Break Secret Codes

Name _____ Date_____

4. b. Find the distances between the occurrences you found and record
        them in the table. Then, for each distance in the table, tell
        whether 3 is a factor.

Repeated Strings in Grandfather's Message

String that repeats	Distance between occurrences	Is 3 a factor?
VNNS	162	yes
SGIAV	105	
GZS (5 times)	30	
	90	
	24	
	51	
GGG	76	
SYR	162	
HUW (4 times)		
IWF (3 times)		
IEW		
GITZ		
ZWA		

c. How did you determine whether 3 was a factor of a number?

    _____

    _____

    _____

d. Do you think 3 is a good guess for the key length of Grandfather's
    message? Why or why not?

    _____

    _____

    _____

© 2006 A K Peters, Ltd., Wellesley, MA

The Cryptoclub: Using Mathematics to Make and Break Secret Codes

(Text page 93)

5. Decrypt Grandfather's message. To save time, use the information in the table to help choose how to turn each Caesar wheel. (It is a long message so you might want to share the work.)

Most Common Letters for Each Wheel

	Most common letters
Wheel 1	W, G, Z, J
Wheel 2	S, W, H, I
Wheel 3	G, A, R, V
English	e, t, a, i

Keyword: _____ _____ _____

Grandfather's Message

[1] A VNNS SGIAV GVDJRJ! WG OOF AB
    1 2 3 1 2 3 1 2 3 1 2 3 1 2 3 1 2 3 1 2 3 1 2

[2] GZS UAZYK PRZWAV HUW HESRVFU
    3 1 2 3 1 2 3 1 2 3 1 2 3 1 2 3 1 2 3 1 2 3 1 2

[3] CGGG GB GZS AADVYCA JWIWF NL
    3 1 2 3 1 2 3 1 2 3 1 2 3 1 2 3 1 2 3 1 2 3 1

[4] HUW BBJHUWFA LWC GT YSYR
    2 3 1 2 3 1 2 3 1 2 3 1 2 3 1 2 3 1 2 3

[5] KICWFVGF. JZWYW VVCWAY, W SGIAV
    1 2 3 1 2 3 1 2 3 1 2 3 1 2 3 1 2 3 1 2 3 1 2 3 1

[6] GBES FZWAQ GGGBRK. ZNLSE A
    2 3 1 2 3 1 2 3 1 2 3 1 2 3 1 2 3 1 2 3 1

(continued)

Chapter 10: Using Common Factors to Crack Vigenère Ciphers

© 2006 A K Peters, Ltd., Wellesley, MA

The Cryptoclub: Using Mathematics to Make and Break Secret Codes

(Text page 93)

[7]

P	E	G	I	T	Z	H		G	Z	S	Z		L	C		N		E	S	G	S	Z		R	P	D	R	J	H
2	3	1	2	3	1	2		3	1	2	3		1	2		3		1	2	3	1	2		3	1	2	3	1	2

[8]

G	G		V	N	N	S		G	Z	S	Z		S	D	C	J	O	V	K	S	Q.		K	I	E	W
3	1		2	3	1	2		3	1	2	3		1	2	3	1	2	3	1	2	3		1	2	3	1

[9]

S	A	G	I	T	Z,		H	U	W	M		N	J	S		F	A	Z	I	W	F	-	V	F		O
2	3	1	2	3	1		2	3	1	2		3	1	2		3	1	2	3	1	2		3	1		2

[10]

I	W	F	L		H	I	E	W		T	B	J	A.		G	Z	S	E	W		A	H	K	H		O	W
3	1	2	3		1	2	3	1		2	3	1	2		3	1	2	3	1		2	3	1	2		3	1

[11]

A	B	J	S	-	V		O	W	Y	D		F	R	L	I	E	F		O	A	V		G	G	S	Y	R		S
2	3	1	2		3		1	2	3	1		2	3	1	2	3	1		2	3	1		2	3	1	2	3		1

[12]

Q	Y	S	W	Z	.
2	3	1	2	3	

© 2006 A K Peters, Ltd., Wellesley, MA

(Text page 94)

6. Below is Grandfather's message encrypted with a different keyword. The goal is to find the key length. (You don't have to decrypt the message since you already know it.)

    a. Underline at least three pairs of repeated strings, including at least two whose distances are not in the table. Then find the distances between occurrences of those strings.

O VLYK TZXTR DLRJPU! OH HDY WY WNS SLRZD EKVTQJ HSH ZFLGOBR

SUGE RT HSH TWALMCY UOJPU GH EKK BZUZVPUT HTS UT WDQS

DXVSCLUF. HKOZP KOYTQM, W QRABO VUAP VNWYB YHZQKG. WDZSC

L HFZXMVE WNSX WU O XHZOW HDDPUZ HZ KGJP WNSX DVDCDOGPG.

YICH KBZXMV, EKKM LUK GTOBSC-LT O GHXM AXXS QRXA. EKKFP

PAGE EK AZUK-W HLRZ CHZICQ GBO VZOVH G QWDOA.

    b. Factor the distances given in the table.

Strings that repeat in message 6	Distance between occurrences
JPU	$52 = 2^2 \times 13$
WNS (3 times)	120
HSH	
EKK (3 times)	120
WNSX	24
ZXMV	
LRZ	208

    c. Make a reasonable guess about what the length of the keyword might be. _____ Explain why your answer is reasonable.

    _____

    _____

Chapter 10: Using Common Factors to Crack Vigenère Ciphers

© 2006 A K Peters, Ltd., Wellesley, MA

The Cryptoclub: Using Mathematics to Make and Break Secret Codes

© 2006 A K Peters, Ltd., Wellesley, MA

The Cryptoclub: Using Mathematics to Make and Break Secret Codes

Name _____ Date _____

(Text pages 94–95)

7. Grandfather's message is encrypted again with a different keyword. (You don't have to decrypt.)

    a. Underline at least three pairs of repeated strings, including at least two whose distances are not in the table. Then find the distances between occurrences of those strings.

A LCMI YGYPU WBDZGI! MM OEU ZR MZI JZPEK FGYMGV XJV XKSHKEK

IGWV FR MZI PZTBYSP IMOWV CK XAW RQIXAWVP KMI GJ NROX

KYRVVBGV. YYMEW LKBMGY, M HFYGV WQDI LZMPP WMGRGJ. PTLIT

Z FKGYIYX MZIO KS T EIVRP XPTGIX MG LCMI MZIO RTIJEKJIW.

KYTV IGGYIY, XAWC CII LAPXVV—BF E XVVR HYTV JHJQ. VYIKW

QWJX UW QQII—B OMNC VXLYTE EGV WVROX S GNRMF.

    b. Factor the distances given in the table.

Strings that repeat in message 7	Distance between occurrences
LCMI	$162 = 2 \times 3^4$
MZI (4 times)	90
XAW (3 times)	114
ROX	162
MZIO	
YTV	
GYIY	
XVV	

    c. Make a reasonable guess about what the length of the keyword might be. _____ Explain why your answer is reasonable.

    _____

    _____

    _____

(Text pages 94–95)

8. Here is Grandfather's message again, encrypted with a different keyword.

    a. Underline at least three pairs of repeated strings, including at least two whose distances are not in the table. Then find the distances between occurrences of those strings.

    I WPGI FDJYH SXAGIR! XI HES XC ELE WXWPS QTSMNS ISI TGPOMNV

    EZWT DC ELE CXAMGDC CMVTG LX TWT YSRIWPVN IXA SF APVI

    SJEPVIDG. HLIAT SMKXCR, M FDJYH SDBP WHXCJ WTDCPW. LPIPV

    I QGZYGWI ELEB IZ E MTILP EMEPVT ID SEVT ISIM PEAVAXHPH.

    SJGP INDJRL, TWTJ ERT HTPVTG—TR A KTCC PJGP JOGB. ELEGT

    XYSI QP QOGT—T AIAA CITJGY ENS HEEKT P NPAXB.

    b. Factor the distances given in the table.

Strings that repeat in message 8	Distance between occurrences
FDJYH	105 = 3 × 5 × 7
ELE (4 times)	
	90
ISI	130
VTG	120
TWT	
PVI	135
PVT	
JGP	
EPV	

    c. Make a reasonable guess about what the length of the keyword might be. _____ Explain why your answer is reasonable.

    _____

    _____

    _____

Chapter 10: Using Common Factors to Crack Vigenère Ciphers

© 2006 A K Peters, Ltd., Wellesley, MA

The Cryptoclub: Using Mathematics to Make and Break Secret Codes

Continue to the next page for Problem 9.

© 2006 A K Peters, Ltd., Wellesley, MA

The Cryptoclub: Using Mathematics to Make and Break Secret Codes

Name _____ **Date** _____

(Text page 96)

9. The following message is encrypted with a Vigenère cipher. Collect data to guess the key, then crack the message. Use the suggestions in 9a–e on the following pages to share the work with your class.

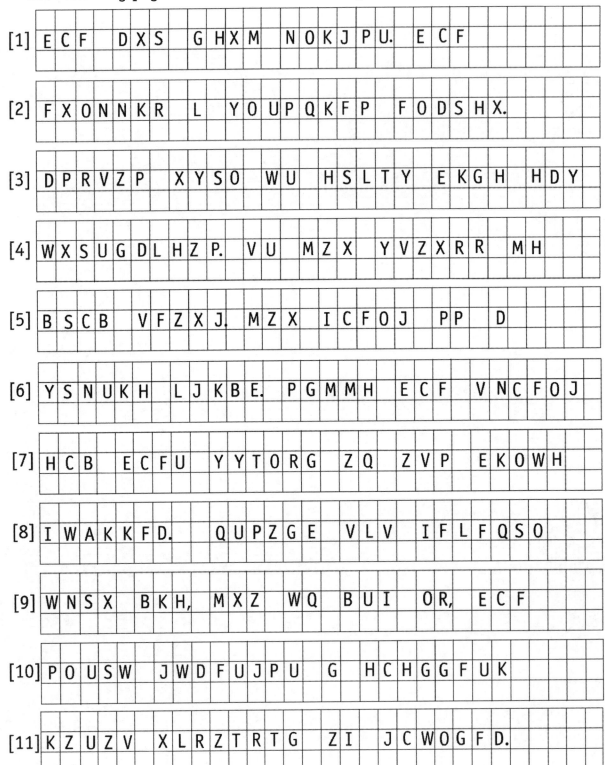

[1] E C F D X S G H X M N O K J P U. E C F

[2] F X O N N K R L Y O U P Q K F P F O D S H X.

[3] D P R V Z P X Y S O W U H S L T Y E K G H H D Y

[4] W X S U G D L H Z P. V U M Z X Y V Z X R R M H

[5] B S C B V F Z X J. M Z X I C F O J P P D

[6] Y S N U K H L J K B E. P G M M H E C F V N C F O J

[7] H C B E C F U Y Y T O R G Z Q Z V P E K O W H

[8] I W A K K F D. Q U P Z G E V L V I F L F Q S O

[9] W N S X B K H, M X Z W Q B U I O R, E C F

[10] P O U S W J W D F U J P U G H C H G G F U K

[11] K Z U Z V X L R Z T R T G Z I J C W O G F D.

© 2006 A K Peters, Ltd., Wellesley, MA

The Cryptoclub: Using Mathematics to Make and Break Secret Codes

Name _____ **Date** _____

9. a. Find strings that repeat. Then find and factor the distances between them. Share your data with your class.

String	Distance	Factorization

b. What is a likely key length? _____ Write numbers under the message to show letters for each wheel.

c. Your teacher will divide the class into groups and assign your group one of the wheels. Find the frequency of the letters for your wheel.

WHEEL # _____		WHEEL # _____	
Letter	Frequency tallies	Letter	Frequency tallies
A		N	
B		O	
C		P	
D		Q	
E		R	
F		S	
G		T	
H		U	
I		V	
J		W	
K		X	
L		Y	
M		Z	

© 2006 A K Peters, Ltd., Wellesley, MA

(Text page 96)

9. d. Record class data about the most common letters for each wheel.

Wheel #	Most common letters (list 3 or 4)

    e. Your teacher will assign your group a few lines of the message to decrypt. Share your decrypted lines with the class. What keyword did you use? _____

10. Describe in your own words how to crack a Vigenère cipher when you do not know anything about the keyword.

_____

_____

_____

_____

_____

_____

_____

_____

_____

_____

© 2006 A K Peters, Ltd., Wellesley, MA

The Cryptoclub: Using Mathematics to Make and Break Secret Codes

Name _____ Date _____

Chapter 11: Introduction to Modular Arithmetic
(Text page 104)

1. Lilah had a play rehearsal that started at 11:00 on Saturday morning. The rehearsal lasted 3 hours. What time did it end? _____

2. Peter was traveling with his family to visit their grandmother and their cousins, Marla and Bethany, near Pittsburgh. The car trip would take 13 hours. If they left at 8:00 AM, what time would they arrive in Pittsburgh? _____

3. The trip to visit their other grandmother takes much longer. First they drive for 12 hours, then stop at a hotel and sleep for about 8 hours. Then they drive about 13 hours more. If they leave at 10:00 AM on Saturday, when will they get to their grandmother's house? _____

4. Use clock arithmetic to solve the following:

   a. 5 + 10 = _____

   b. 8 + 11 = _____

   c. 7 + 3 = _____

   d. 9 + 8 + 8 = _____

5. Jenny's family is planning a 5-hour car trip. They want to arrive at 2 PM. What time should they leave? _____

6. In Problem 5, we moved backward around the clock. This is the same as subtracting in clock arithmetic. Solve the following subtraction problems using clock arithmetic. Use the clock, if you like, to help you:

   a. 3 – 7 = _____

   b. 5 – 6 = _____

   c. 2 – 3 = _____

   d. 5 – 10 = _____

© 2006 A K Peters, Ltd., Wellesley, MA

Name _____ **Date** _____

(Text page 106)

7. Write the following 12-hour times using the 24-hour system:

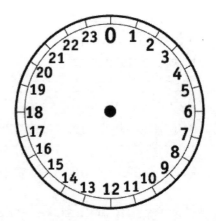

    a. 3 PM   _____

    b. 9 AM   _____

    c. 11:15 PM _____

    d. 4:30 AM _____

    e. 6:45 PM _____

    f. 8:30 PM _____

8. Write the following 24-hour times as 12-hour times, using AM or PM.

    a. 13:00 _____

    b. 5:00 _____

    c. 19:15 _____

    d. 21:00 _____

    e. 11:45 _____

    f. 15:30 _____

9. Use clock arithmetic on a 24-hour clock to solve the following:

    a. 20 + 6 = _____   b. 11 + 17 = _____

    c. 22 – 8 = _____   d. 8 – 12 = _____

© 2006 A K Peters, Ltd., Wellesley, MA

The Cryptoclub: Using Mathematics to Make and Break Secret Codes

© 2006 A K Peters, Ltd., Wellesley, MA

Name _____ Date _____

(Text page 106)

10. Solve the following on a 10-hour clock:

a. 8 + 4 = _____

b. 5 + 8 = _____

c. 7 + 7 = _____

d. 10 + 15 = _____

e. 6 − 8 = _____

f. 3 + 5 = _____

11. Challenge: Is 2 + 2 always 4? Find a clock for which this is not true.

_____

_____

_____

_____

_____

_____

The Cryptoclub: Using Mathematics to Make and Break Secret Codes

Name _____ Date_____

12. The figure shows numbers wrapped around a 12-hour clock.

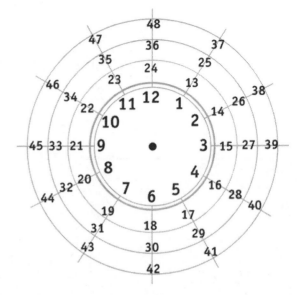

   a. List all numbers between 1 and 48 that have the same position on a 12-hour clock as 3.

   _____

   b. If the number wrapping continues, what numbers between 49 and 72 would have the same position on a 12-hour clock as 3?

   _____

13. a. List all numbers between 1 and 48 that have the same position on a 12-hour clock as 8.

   _____

   b. If the number wrapping continues, what numbers between 49 and 72 would have the same position on a 12-hour clock as 8?

   _____

14. a. How can you use arithmetic to describe numbers that have the same position on a 12-hour clock as 5?

   _____

   _____

   b. What numbers between 49 and 72 have the same position on the 12-hour clock as 5?

   _____

© 2006 A K Peters, Ltd., Wellesley, MA

The Cryptoclub: Using Mathematics to Make and Break Secret Codes

© 2006 A K Peters, Ltd., Wellesley, MA

The Cryptoclub: Using Mathematics to Make and Break Secret Codes

Name _____ Date_____

(Text page 108)

15. List three numbers equivalent to each number.

    a. 6 mod 12 _____

    b. 9 mod 12 _____

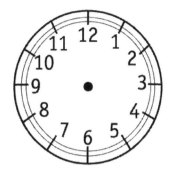

16. List three numbers equivalent to each number.

    a. 2 mod 10 _____

    b. 9 mod 10 _____

    c. 0 mod 10 _____

17. List three numbers equivalent to each number.

    a. 1 mod 5 _____

    b. 3 mod 5 _____

    c. 2 mod 5 _____

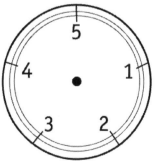

© 2006 A K Peters, Ltd., Wellesley, MA

Name _____ **Date**_____

(Text page 110)

18. Reduce each number.

   a. 8 mod 5 = _____

   b. 13 mod 5 =_____

   c. 6 mod 5 = _____

   d. 4 mod 5 = _____

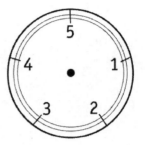

19. Reduce each number.

   a. 18 mod 12 = _____   b. 26 mod 12 =_____

   c. 36 mod 12 = _____   d. 8 mod 12 = _____

20. Reduce each number.

   a. 8 mod 3 = _____   b. 13 mod 6 = _____

   c. 16 mod 11 = _____   d. 22 mod 7 = _____

The Cryptoclub: Using Mathematics to Make and Break Secret Codes

© 2006 A K Peters, Ltd., Wellesley, MA

The Cryptoclub: Using Mathematics to Make and Break Secret Codes

Name_____ Date_____

(Text page 110)

21. Reduce each number.

    a. −4 mod 12 =_____

    b. −1 mod 12 = _____

    c. −6 mod 12 =_____

    d. −2 mod 12 = _____

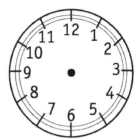

22. Reduce each number.

    a. −4 mod 10 =_____

    b. −1 mod 10 = _____

    c. −6 mod 10 =_____

    d. −2 mod 10 = _____

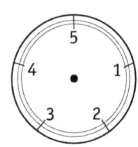

23. Reduce each number.

    a. −3 mod 5 = _____

    b. −1 mod 5 = _____

    c. 8 mod 5 = _____

    d. 7 mod 5 = _____

24. Reduce each number.

    a. −2 mod 24 =_____

    b. 23 mod 20 = _____

    c. 16 mod 11 = _____

    d. −3 mod 20 = _____

Continue to the next chapter.

© 2006 A K Peters, Ltd., Wellesley, MA

© 2006 A K Peters, Ltd., Wellesley, MA

The Cryptoclub: Using Mathematics to Make and Break Secret Codes

Name _____ **Date** _____

Chapter 12: More Modular Arithmetic
(Text pages 115–116)

1. Reduce the following numbers mod 26:

   a. 29 = _____ mod 26

   b. 33 = _____ mod 26

   c. 12 = _____ mod 26

   d. 40 = _____ mod 26

   e. −4 = _____ mod 26

   f. 52 = _____ mod 26

   g. −10 = _____ mod 26

   h. −7 = _____ mod 26

2. Encrypt the name "Jack" using the times-5 cipher. The first two letters are done for you. The rule for encrypting is given in the table.

Times-5 Cipher	J	a	c	k
change letters to numbers (use cipher strip)	9	0		
multiply by 5	45	0		
reduce mod 26	19	0		
change numbers to letters	T	A		

3. Encrypt "cryptography" using a times-3 cipher. The first two letters are done for you.

Times-3 Cipher	c	r	y	p	t	o	g	r	a	p	h	y
change letters to numbers (use cipher strip)	2	17										
multiply by 3	6	51										
reduce mod 26	6	25										
change numbers to letters	G	Z										

(Text pages 117–118)

4. Reduce each number.

    a. 175 mod 26 = _____   b. 106 mod 26 = _____

    c. 78 mod 26 = _____   d. 150 mod 26 = _____

5. Reduce each number. (Hint: Try subtracting multiples of 26 such as 10 × 26 = 260.)

    a. 586 mod 26 = _____   b. 792 mod 26 = _____

    c. 541 mod 26 = _____   d. 364 mod 26 = _____

Return to Text

6. Use a calculator to help you reduce each number.

    a. 254 mod 24 = _____   b. 500 mod 5 = _____

    c. 827 mod 26 = _____   d. 1500 mod 26 = _____

    e. 700 mod 9 = _____   f. 120 mod 11 = _____

© 2006 A K Peters, Ltd., Wellesley, MA

The Cryptoclub: Using Mathematics to Make and Break Secret Codes

(Text pages 118–119)

7. Reduce each number.

    a. 500 mod 7 = _____

    b. 1000 mod 24 = _____

    c. 25,000 mod 5280 = _____

    d. 10,000 mod 365 = _____

8. Choose one of the numbers you reduced in Problem 6. Write how you would explain to a friend the way you reduced your number.

_____

_____

_____

_____

_____

_____

Return to Text

9. Encrypt "trick," using a times-11 cipher. Use Tim's shortcut when it makes your work easier.

Times-11 Cipher	z	e	s	t	y
change letters to numbers					
multiply by 11					
reduce mod 26					
change numbers to letters					

© 2006 A K Peters, Ltd., Wellesley, MA

(Text page 120)

10. Astronauts left on a Sunday for a mission into space. On what day of the week would they return if they were gone for

    a. 4 days? _____

    b. 15 days? _____

    c. 100 days? _____

    d. 1000 days? _____

11. If today is Wednesday, what day of the week will it be in

    a. 3 days? _____

    b. 75 days? _____

    c. 300 days? _____

Leap Years. There are 365 days in a year, except for leap years. In a leap year, an extra day (February 29) is added, making 366 days. Leap years occur in years divisible by 4, except at the beginning of some centuries. Years that begin new centuries are not leap years unless they are divisible by 400. So 1900 was not a leap year but 2000 was.

12.   a. 2004 was a leap year. What are the next two leap years? _____

        b. Which of the following century years are leap years? _____
            1800, 2100, 2400

        c. Which of the following years were leap years? _____
            1996, 1776, 1890

© 2006 A K Peters, Ltd., Wellesley, MA

The Cryptoclub: Using Mathematics to Make and Break Secret Codes

(Text page 120)

13. If the Fourth of July is on Tuesday this year, on what day of the week will it be next year? (Assume that next year is not a leap year.) Explain how you got your answer.

_____

_____

_____

_____

_____

14. a. What is today's day and date? _____

 b. What day of the week will it be on today's date next year? Your answer will depend on whether or not a leap year is involved. Explain how you got your answer.

_____

_____

_____

_____

_____

© 2006 A K Peters, Ltd., Wellesley, MA

The Cryptoclub: Using Mathematics to Make and Break Secret Codes

(Text page 120)

15. a. On what day and date will your next birthday be? (You may use a calendar.)

_____

_____

   b. On what day of the week will your twenty-first birthday be? Answer without using a calendar. Don't forget about leap years. Explain how you got your answer.

_____

_____

_____

_____

_____

_____

_____

_____

© 2006 A K Peters, Ltd., Wellesley, MA

The Cryptoclub: Using Mathematics to Make and Break Secret Codes

The Cryptoclub: Using Mathematics to Make and Break Secret Codes

© 2006 A K Peters, Ltd., Wellesley, MA

Name —————————————————————— **Date** ——————————

Chapter 13: Multiplicative Ciphers
(Text page 126)

1. a. Complete the times-3 cipher table. (Tip: You can use patterns such as multiplying by 3s to multiply quickly.)

plaintext:	a	b	c	d	e	f	g	h	i	j	k	l	m	n	o	p	q	r	s	t	u	v	w	x	y	z
numbers:	0	1	2	3	4	5	6	7	8	9	10	11	12	13	14	15	16	17	18	19	20	21	22	23	24	25
× 3 (mod 26):	0	3	6	9	12	15	18	21	24	1	4	7														
ciphertext:	A	D	G	J	M	P	S	V	Y	B	E	H														

b. Decrypt the following message Evie wrote using the times-3 cipher.

JAN, Y ENQO OVAF UQI OZQFM.

c. **Riddle:** What has one foot on each end and one foot in the middle? (It was encrypted using the times-3 cipher.)

Answer: A UAZJCFYGE

(Text page 126)

2.

a. Complete the times-2 cipher table.

plaintext:	a	b	c	d	e	f	g	h	i	j	k	l	m	n	o	p	q	r	s	t	u	v	w	x	y	z
numbers:	0	1	2	3	4	5	6	7	8	9	10	11	12	13	14	15	16	17	18	19	20	21	22	23	24	25
× 2 (mod 26):				6							20				2							16				
ciphertext:				G							U				C							Q				

b. Use the times-2 cipher to encrypt the words **ant** and **nag**. Is there anything unusual about your answers?

   a n t

   n a g

c. Make a list of pairs of letters that are encrypted the same way using a times-2 cipher. For example, **a** and **n** are both encrypted as **A**, **b** and **o** are both encrypted as **C**.

d. Make a list of several pairs of words that are encrypted the same way using the times-2 cipher.

e. Decrypt **KOI** in more than one way to get different English words.

f. Does multiplying by 2 give a good cipher? Why or why not?

© 2006 A K Peters, Ltd., Wellesley, MA

© 2006 A K Peters, Ltd., Wellesley, MA

Name ——————————————————————— **Date** ——————

(Text pages 126–127)

3. a. Complete the times-5 cipher table. Then decrypt the quotations.

plaintext:	a	b	c	d	e	f	g	h	i	j	k	l	m	n	o	p	q	r	s	t	u	v	w	x	y	z
numbers:	0	1	2	3	4	5	6	7	8	9	10	11	12	13	14	15	16	17	18	19	20	21	22	23	24	25
× 5 (mod 26):	0	5	10				4																			
ciphertext:	A	F	K				E																			

b. FU MWHU QSW XWR QSWH ZUUR ON RJU HOEJR

XDAKU, RJUN MRANP ZOHI.

—Abraham Lincoln

c. RJU OIXSHRANR RJONE OM NSR RS MRSX CWUMROSNONE.

—Albert Einstein

(Text page 127)

4.   a.   Complete the times-13 cipher table.

plaintext:	a	b	c	d	e	f	g	h	i	j	k	l	m	n	o	p	q	r	s	t	u	v	w	x	y	z
numbers:	0	1	2	3	4	5	6	7	8	9	10	11	12	13	14	15	16	17	18	19	20	21	22	23	24	25
× 13 (mod 26):																										
ciphertext:																										

b.   Encrypt using the times-13 cipher:

i n p u t

a l t e r

c.   Does multiplying by 13 give a good cipher?   Why or why not?

_____

_____

_____

© 2006 A K Peters, Ltd., Wellesley, MA

The Cryptoclub: Using Mathematics to Make and Break Secret Codes

Name —————————————————————— **Date** ————————————

© 2006 A K Peters, Ltd., Wellesley, MA

The Cryptoclub: Using Mathematics to Make and Break Secret Codes

Class Activity (Text page 127)

a. Choose one even number and one odd number between 4 and 25 to investigate. One number should be big, the other small. (Groups that finish early can work on the numbers not yet chosen.) ——————

b. Compute cipher tables using your numbers as multiplicative keys. Decide which of your numbers make good multiplicative keys (that is, which numbers encrypt every letter differently). (Use the extra tables on the back of this page if you compute more than two ciphers.)

Times- ——————— Cipher

plaintext:	a	b	c	d	e	f	g	h	i	j	k	l	m	n	o	p	q	r	s	t	u	v	w	x	y	z
numbers:	0	1	2	3	4	5	6	7	8	9	10	11	12	13	14	15	16	17	18	19	20	21	22	23	24	25
x—— (mod 26):																										
ciphertext:																										

Good key or bad key? ——————

Times- ——————— Cipher

plaintext:	a	b	c	d	e	f	g	h	i	j	k	l	m	n	o	p	q	r	s	t	u	v	w	x	y	z
numbers:	0	1	2	3	4	5	6	7	8	9	10	11	12	13	14	15	16	17	18	19	20	21	22	23	24	25
x—— (mod 26):																										
ciphertext:																										

Good key or bad key? ——————

c. Pool your information with the rest of the class. Describe a pattern that tells which numbers give good keys.

————————————————————————————————

Extra Tables:

Times-_____ Cipher

Good key or bad key? _____

plaintext:	a	b	c	d	e	f	g	h	i	j	k	l	m	n	o	p	q	r	s	t	u	v	w	x	y	z
numbers:	0	1	2	3	4	5	6	7	8	9	10	11	12	13	14	15	16	17	18	19	20	21	22	23	24	25
×_____ (mod 26):																										
ciphertext:																										

Times-_____ Cipher

Good key or bad key? _____

plaintext:	a	b	c	d	e	f	g	h	i	j	k	l	m	n	o	p	q	r	s	t	u	v	w	x	y	z
numbers:	0	1	2	3	4	5	6	7	8	9	10	11	12	13	14	15	16	17	18	19	20	21	22	23	24	25
×_____ (mod 26):																										
ciphertext:																										

Times-_____ Cipher

Good key or bad key? _____

plaintext:	a	b	c	d	e	f	g	h	i	j	k	l	m	n	o	p	q	r	s	t	u	v	w	x	y	z
numbers:	0	1	2	3	4	5	6	7	8	9	10	11	12	13	14	15	16	17	18	19	20	21	22	23	24	25
×_____ (mod 26):																										
ciphertext:																										

© 2006 A K Peters, Ltd., Wellesley, MA

The Cryptoclub: Using Mathematics to Make and Break Secret Codes

© 2006 A K Peters, Ltd., Wellesley, MA

Name _____ Date _____

(Text pages 129–130)

5. Which of the following pairs of numbers are relatively prime?

    a. 3 and 12 _____   b. 13 and 26 _____

    c. 10 and 21 _____   d. 15 and 22 _____

    e. 8 and 20 _____   f. 2 and 14 _____

6.   a. List 3 numbers that are relatively prime to 26.

    _____

    b. List 3 numbers that are relatively prime to 24.

    _____

7. Which numbers make good multiplicative keys for each of the following alphabets?

    a. Russian; 33 letters _____

    b. Lilah's "alphabet", which consists of the 26 English letters and the period, comma, question mark, and blank space _____

    _____

    c. Korean; 24 letters _____

    d. Arabic; 28 letters. This alphabet is used to write about 100 languages, including Arabic, Kurdish, Persian, and Urdu (the main language of Pakistan). _____

    e. Portuguese; 23 letters _____

The Cryptoclub: Using Mathematics to Make and Break Secret Codes

(Text page 130)

8. Compute the table for each cipher, then decrypt the quote:

   a. Times-7 cipher

plaintext:	a	b	c	d	e	f	g	h	i	j	k	l	m	n	o	p	q	r	s	t	u	v	w	x	y	z
numbers:	0	1	2	3	4	5	6	7	8	9	10	11	12	13	14	15	16	17	18	19	20	21	22	23	24	25
× 7 (mod 26):																										
ciphertext:																										

UKP OXAPAODCP EW YXAD YC VU YXCN YC DXENS

NU UNC EW ZUUSENQ.

—H. Jackson Brown, Jr.

© 2006 A K Peters, Ltd., Wellesley, MA

The Cryptoclub: Using Mathematics to Make and Break Secret Codes

The Cryptoclub: Using Mathematics to Make and Break Secret Codes

© 2006 A K Peters, Ltd., Wellesley, MA

(Text page 130)

8. b. Times-9 cipher

plaintext:	a	b	c	d	e	f	g	h	i	j	k	l	m	n	o	p	q	r	s	t	u	v	w	x	y	z
numbers:	0	1	2	3	4	5	6	7	8	9	10	11	12	13	14	15	16	17	18	19	20	21	22	23	24	25
× 9 (mod 26):																										
ciphertext:																										

PLK EWGP KZLAYGPUNC PLUNC UN VUTK UG JKUNC

UNGUNSKXK.

—Anne Morrow Lindbergh

(Text page 130)

8. c.   Times-11 cipher

plaintext:	a	b	c	d	e	f	g	h	i	j	k	l	m	n	o	p	q	r	s	t	u	v	w	x	y	z
numbers:	0	1	2	3	4	5	6	7	8	9	10	11	12	13	14	15	16	17	18	19	20	21	22	23	24	25
× 11 (mod 26):																										
ciphertext:																										

IS GNYI IZAB IS AFS, LMB NYB IZAB IS CAE LS.

—William Shakespeare

© 2006 A K Peters, Ltd., Wellesley, MA

© 2006 A K Peters, Ltd., Wellesley, MA

(Text page 130)

8.   d.   Times-25 cipher (Hint: 25 ≡ −1 (mod 26).)

plaintext:	a	b	c	d	e	f	g	h	i	j	k	l	m	n	o	p	q	r	s	t	u	v	w	x	y	z
numbers:	0	1	2	3	4	5	6	7	8	9	10	11	12	13	14	15	16	17	18	19	20	21	22	23	24	25
x −1 (mod 26):																										
ciphertext:																										

HTW ZWUSNNSNU SI HTW OMIH

SOLMJHANH

LAJH MV HTW EMJQ.

—Plato

(Text page 131)

9.   Look at your cipher tables from Problem 8.

   a.   How was **a** encrypted? _____
   Will this be the same in all multiplicative ciphers? Give a reason for
   your answer.

   _____

   _____

   _____

   _____

   b.   How was **n** encrypted? _____
   Challenge. Show that this will be the same in all multiplicative
   ciphers. Hint: Since all multiplicative keys are odd numbers, every
   key can be written as an even number plus 1.

   _____

   _____

   _____

   _____

© 2006 A K Peters, Ltd., Wellesley, MA

The Cryptoclub: Using Mathematics to Make and Break Secret Codes

© 2006 A K Peters, Ltd., Wellesley, MA

Name _____   Date _____

Chapter 14: Using Inverses to Decrypt
(Text pages 135–136)

1. Compute the following in regular arithmetic.

   a. $2 \times \frac{1}{2}$ = _____

   b. $\frac{1}{4} \times 4$ = _____

   c. $7 \times \frac{1}{7}$ = _____

2. Complete:

   a. $3 \xrightarrow{\times 2} 6 \xrightarrow{\times \frac{1}{2}}$ _____

   b. $6 \xrightarrow{\times 3} 18 \xrightarrow{\times \frac{1}{3}}$ _____

   c. $2 \xrightarrow{\times 5} 10 \xrightarrow{\times \boxed{}}$ __2__   (fill in the box)

   d. $4 \xrightarrow{\times 6} 24 \xrightarrow{\times \boxed{}}$ __4__   (fill in the box)

Return to Text

3. Test Abby's theory that if you multiply by 3 and then by 9 (and reduce mod 26) you get back what you started with:

   a. $6 \xrightarrow{\times 3} 18 \xrightarrow{\times 9} 162 \equiv$ _____ (mod 26)

   b. $2 \xrightarrow{\times 3}$ _____ $\xrightarrow{\times 9}$ _____ $\equiv$ _____ (mod 26)

   c. $10 \xrightarrow{\times 3}$ _____ $\xrightarrow{\times 9}$ _____ $\equiv$ _____ (mod 26)

(Text page 136)

4. Where was Tim's second treasure hidden? Finish decryping his clue to find out. (He used a multiplicative cipher with key 3).

	plaintext	t								
	reduce mod 26	19								
	multiply by inverse	45								
	change letters to numbers	5								
	ciphertext	F	Q	T		C	V	M	H	P

© 2006 A K Peters, Ltd., Wellesley, MA

The Cryptoclub: Using Mathematics to Make and Break Secret Codes

© 2006 A K Peters, Ltd., Wellesley, MA

Name _____ Date _____

(Text page 137)

5. Look at the tables of multiplicative ciphers you have already worked out. Find the column that has 1 in the product row and use this to find other pairs of numbers that are inverses mod 26. Save these for later.

_____

Return to Text

6. Since $5 \times 21 = 105 \equiv 1 \pmod{26}$, 5 and 21 are inverses of each other (mod 26). Find another way to factor 105. Use this to find another pair of mod 26 inverses.

_____

7. The following was encrypted by multiplying by 21. Decrypt. (Hint: See Problem 6 for the inverse of 21.)

plaintext	a																	
reduce mod 26	0																	
multiply by inverse	0																	
change letters to numbers	0																	
ciphertext	A		U	M	X	X		B	M	N	L	O		A		U	A	K.

—Orison Swett Marden

(Text page 138)

8. Find another inverse pair by looking at the negatives of the inverses we have already found.

9. What is the inverse of 25 mod 26? (Hint: $25 \equiv -1 \pmod{26}$.)

© 2006 A K Peters, Ltd., Wellesley, MA

The Cryptoclub: Using Mathematics to Make and Break Secret Codes

© 2006 A K Peters, Ltd., Wellesley, MA

Name _____ Date _____

(Text page 138)

10. a. Make a list of all the pairs of inverses you and your classmates have found. (Keep this list to help you decrypt messages.)

_____

_____

b. What numbers are not on your list? (Not all numbers have inverses mod 26.)

_____

c. Describe a pattern that tells which numbers between 1 and 25 have inverses mod 26.

_____

_____

11. **Challenge.** Explain why even numbers do not have inverses mod 26.

_____

_____

_____

_____

_____

_____

_____

The Cryptoclub: Using Mathematics to Make and Break Secret Codes

(Text page 139)

Solve these problems by multiplying by the inverse.

12. **Riddle:** What word is pronounced wrong by the best of scholars?
    Answer (encrypted with a times-9 cipher):

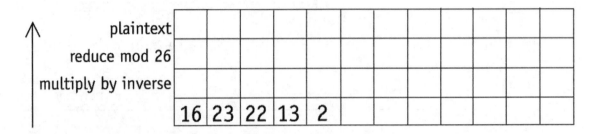

13. **Riddle:** What's the best way to catch a squirrel?
    Answer (encrypted with a times-15 cipher):

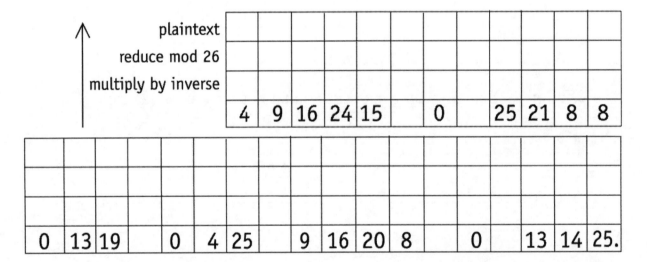

© 2006 A K Peters, Ltd., Wellesley, MA

The Cryptoclub: Using Mathematics to Make and Break Secret Codes

Chapter 14: Using Inverses to Decrypt

© 2006 A K Peters, Ltd., Wellesley, MA

Name _____ Date _____

(Text page 139)

14. **Challenge.** Investigate inverses for one of the alphabets listed below. Find all pairs of numbers that are inverses of each other.

   a. Russian; 33 letters
   b. The English alphabet, and the period, comma, question mark, and blank space; 30 "letters"
   c. Korean; 24 letters (Note: There is something unusual about the inverses for this alphabet.)

The Cryptoclub: Using Mathematics to Make and Break Secret Codes

(Text page 141)

15. Where did Evie's note say to meet? Finish decrypting to find out. Show
    your work below the message.

```
                  me e t  me   a t   the          a
                  IUUR  IU  AR  RJU  DOFHAHQ
```

© 2006 A K Peters, Ltd., Wellesley, MA

The Cryptoclub: Using Mathematics to Make and Break Secret Codes

© 2006 A K Peters, Ltd., Wellesley, MA

Name _____ Date _____

(Text page 141)

16. The following messages were encrypted with multiplicative ciphers. A few
    letters in each message have been decrypted. For each message, write an
    equivalence that involves the key. Then solve the equivalence to find the
    key. Use the inverse of the key to help decrypt. Show your work below
    the messages. (Note: No table is given, so you decide how you want to
    organize your work.)

```
        i t e   i t                    e    t    t    e  e
a.  QXUPK  UP  WN  IWYX  LKAXP  PLAP  KHKXI

        i     t e   e t           i     t e     e
    BAI  UG  PLK  JKGP  BAI  UN  PLK  IKAX.
```

—Ralph Waldo Emerson

The Cryptoclub: Using Mathematics to Make and Break Secret Codes

(Text page 141)

```
        the    e t      t    hee         e
16. b. ZBI  PIKZ  SAW  ZC  EBIIV  WCOVKIJX  OR

        t  t    t    hee      e  e  e
       QK  ZC  ZVW  ZC  EBIIV  KCYICNI  IJKI  OR.
```

—Mark Twain

© 2006 A K Peters, Ltd., Wellesley, MA

© 2006 A K Peters, Ltd., Wellesley, MA

Name _____ **Date** _____

(Text page 142)

17. For each of the following, find the most common letters in the message. Use this information or other reasoning to guess a few letters of the message. Then find the encryption key by solving an equivalence. Use the inverse of the key to help decrypt. Show your work below the messages.

a. A HOYYCQCYV YOOY VFO RCLLCUSTVG CN

OPOBG KHHKBVSNCVG; AN KHVCQCYV YOOY

VFO KHHKBVSNCVG CN OPOBG RCLLCUSTVG.

—Winston Churchill

(Text page 142)

17. b. JTGAJ A SAN AO RG MO, ANL RG UMXX

TGSAMN AO RG MO. JTGAJ A SAN AO RG

QIEXL VG, ANL RG UMXX VGQISG URAJ

RG ORIEXL VG.

—Ralph Waldo Emerson

© 2006 A K Peters, Ltd., Wellesley, MA

The Cryptoclub: Using Mathematics to Make and Break Secret Codes

© 2006 A K Peters, Ltd., Wellesley, MA

Name _____ Date_____

Chapter 15: Affine Ciphers
(Text page 145)

1. How many different additive ciphers are possible? That is, how many different numbers can be keys for additive ciphers? Explain how you got your answer.

_____

_____

_____

_____

2. How many different multiplicative ciphers are possible? That is, how many different numbers make good keys for multiplicative ciphers? Explain how you got your answer. (Remember, the good multiplicative keys are those that are relatively prime to 26.)

_____

_____

_____

_____

(Text page 146)

3. Encrypt "secret" using the (3, 7)-affine cipher.

plaintext	s	e	c	r	e	t
change letters to numbers	18					
multiply by 3	54					
add 7	61					
mod 26	9					
ciphertext	J	T				

4. Encrypt "secret" using the (5, 8)-affine cipher.

plaintext	s	e	c	r	e	t
change letters to numbers	18					
multiply by 5						
add 8						
mod 26						
ciphertext	U	C				

© 2006 A K Peters, Ltd., Wellesley, MA

The Cryptoclub: Using Mathematics to Make and Break Secret Codes

© 2006 A K Peters, Ltd., Wellesley, MA

The Cryptoclub: Using Mathematics to Make and Break Secret Codes

(Text pages 146–147)

5. Some affine ciphers are the same as other ciphers we have already explored.

   a. What other cipher is the same as the (3, 0)-affine cipher?

   _____

   b. What other cipher is the same as the (1, 8)-affine cipher?

   _____

6. Suppose that Dan and Tim changed their key to get a different affine cipher each day. Would they have enough ciphers to have one for each day of the year? Explain.

_____

_____

_____

_____

_____

_____

Return to Text

7. **Riddle:** What insects are found in clocks?

   Answer (encrypted with a (3, 7)-affine cipher):

plaintext					
mod 26					
multiply by inverse of 3					
subtract 7					
change letters to numbers					
ciphertext	M	F	N	L	J

8. Decrypt the girls' invitation. It was encrypted with a (5, 2)-affine cipher. There is no table given, so you can decide how you would like to organize your work.

SUY   CJW   QPDQTWR   TU

FQFCL   CPR   HWMAS'O

HWCML   ZCJTS.

KWWT   CT   TLW   ZCDQFQUP

HS   TLW   FCAW   CT   2ZK   UP

OCTYJRCS.

© 2006 A K Peters, Ltd., Wellesley, MA

The Cryptoclub: Using Mathematics to Make and Break Secret Codes

© 2006 A K Peters, Ltd., Wellesley, MA

Name _____ Date _____

(Text page 150)

9. Each of the following was encrypted with an affine cipher. A few letters have been decrypted. For each message, write equivalences involving the encryption key (m, b). Solve the equivalences to find the m and b. Then decrypt the message.

     e   e   e n

a. M C Z R N   H Z Y J W D M I   M P Y A E N   R Y   I C R W I V K

    e   e   n   n e e   n   e   e

   M J M Z K   G C P ' I   P M M D,   L A R   P Y R   M J M Z K

       n   e e

  G C P ' I   E Z M M D.

—Mahatma Gandhi

The Cryptoclub: Using Mathematics to Make and Break Secret Codes

(Text pages 150–151)

        i   a   a   t   a   a   i
9. b. S GY G UKWWMUU PORGQ BMWGKUM S

        a   a   a   i   i   i
     XGR G HZSMTR AXO BMDSMFMR ST YM

        a   i   i   t   a   t   a t t
     GTR S RSRT'P XGFM PXM XMGZP PO

        t   i
  DMP XSY ROAT.

—Abraham Lincoln

© 2006 A K Peters, Ltd., Wellesley, MA

The Cryptoclub: Using Mathematics to Make and Break Secret Codes

© 2006 A K Peters, Ltd., Wellesley, MA

The Cryptoclub: Using Mathematics to Make and Break Secret Codes

Name _____ Date _____

(Text page 151)

10. a. Guess a few letters of Peter and Tim's note. Then solve two
      equivalences to find *m* and *b*. Show your work.
    b. Decrypt Peter and Tim's note.

UO   UYZZ   WAIO   TAA   ERF

NBYRG   E   WEQO.

JOTOB   ERF   TYI

(Text page 151)

11. Each of the following was encrypted with an affine cipher. Use letter frequencies or any other information to figure out a few of the letters. Write equivalences using the letter substitutions. Solve the equivalences to find the key (*m*, *b*). Then decrypt.

a. B O I O I N O B   R A T   A R Z M   T A   K E M   T P O

BYGPT TPYRG YR TPO BYGPT JZEWO,

NCT XEB IABO FYXXYWCZT KTYZZ, TA

ZOELO CRKEYF TPO UBARG TPYRG ET

TPO TOIJTYRG IAIORT

—Benjamin Franklin

© 2006 A K Peters, Ltd., Wellesley, MA

The Cryptoclub: Using Mathematics to Make and Break Secret Codes

© 2006 A K Peters, Ltd., Wellesley, MA

Name _____ Date _____

(Text page 151)

11. b. RY XDP CBEJ BO BSSKJ BOU R CBEJ

BO BSSKJ BOU TJ JIFCBONJ ACJLJ

BSSKJL ACJO XDP BOU R TRKK LARKK

JBFC CBEJ DOJ BSSKJ. QPA RY XDP

CBEJ BO RUJB BOU R CBEJ BO RUJB

BOU TJ JIFCBONJ ACJLJ RUJBL, ACJO

JBFC DY PL TRKK CBEJ ATD RUJBL.

—George Bernard Shaw

The Cryptoclub: Using Mathematics to Make and Break Secret Codes

Continue to the next chapter.

© 2006 A K Peters, Ltd., Wellesley, MA

© 2006 A K Peters, Ltd., Wellesley, MA

The Cryptoclub: Using Mathematics to Make and Break Secret Codes

Name _____ **Date**_____

Chapter 16: Finding Prime Numbers
(Text page 158)

1. Find whether the following are prime numbers. Explain how you know.

a. 343 _____

_____

b. 1019 _____

_____

c. 1369 _____

_____

d. 2417 _____

_____

e. 2573 _____

_____

f. 1007 _____

_____

(Text page 160)

2. Follow the steps for the Sieve of Eratosthenes to find all prime numbers from 1 to 50.

The Sieve of Eratosthenes

a. Cross out 1 since it is not prime.

b. Circle 2 since it is prime. Then cross out all remaining multiples of 2, since they can't be prime. (Why not?)

c. Circle 3, the next prime. Cross out all remaining multiples of 3, since they can't be prime.

d. Circle the next number that hasn't been crossed out. It is prime. (Why?) Cross out all remaining multiples of that number.

e. Repeat Step D until all numbers are either circled or crossed out.

Prime Numbers from 1 to 50

1	2	3	4	5	6	7	8	9	10
11	12	13	14	15	16	17	18	19	20
21	22	23	24	25	26	27	28	29	30
31	32	33	34	35	36	37	38	39	40
41	42	43	44	45	46	47	48	49	50

Primes less than 50: _____

_____

3. As you followed the steps in Problem 2, you probably found that the multiples of the bigger prime numbers had already been crossed out. What was the largest prime whose multiples were not already crossed out by smaller numbers? _____

© 2006 A K Peters, Ltd., Wellesley, MA

The Cryptoclub: Using Mathematics to Make and Break Secret Codes

© 2006 A K Peters, Ltd., Wellesley, MA

The Cryptoclub: Using Mathematics to Make and Break Secret Codes

Name _____ Date _____

(Text page 160)

4. a. Use the Sieve of Eratosthenes to find all primes between 1 and 130.
   Each time you work with a new prime, write in Table 2 the first of its
   multiples not already crossed out by a smaller prime.

Finding a Pattern

Prime	First multiple not yet crossed out
2	4
3	9
5	

← *Example: When the prime is 3, the first multiple to consider is 6, but 6 has already been crossed out. Therefore, 9 is the first multiple of 3 not already crossed out by a smaller prime.*

Finding Primes 1 to 130

1	2	3	4	5	6	7	8	9	10
11	12	13	14	15	16	17	18	19	20
21	22	23	24	25	26	27	28	29	30
31	32	33	34	35	36	37	38	39	40
41	42	43	44	45	46	47	48	49	50
51	52	53	54	55	56	57	58	59	60
61	62	63	64	65	66	67	68	69	70
71	72	73	74	75	76	77	78	79	80
81	82	83	84	85	86	87	88	89	90
91	92	93	94	95	96	97	98	99	100
101	102	103	104	105	106	107	108	109	110
111	112	113	114	115	116	117	118	119	120
121	122	123	124	125	126	127	128	129	130

Primes less than 130: _____

_____

_____

(Text page 160)

4. b. Look at your table from 4a. Describe a pattern that tells, for any prime number, its first multiple not already crossed out by smaller prime numbers. _____

_____

_____

c. When sieving for primes between 1 and 130, what was the largest prime whose multiples were not already crossed out by smaller numbers?

_____

d. After you had crossed out the multiples of enough primes, you could stop because only prime numbers were left. When did this happen?

_____

_____

_____

5. a. Suppose that you used the sieve method to find the primes between 1 and 200. List the primes whose multiples you would have to cross out before only primes were left. Explain why.

_____

_____

_____

b. Suppose that you used the sieve method to find the primes between 1 and 1000. List the primes whose multiples you would have to cross out before only primes were left.

_____

_____

_____

© 2006 A K Peters, Ltd., Wellesley, MA

The Cryptoclub: Using Mathematics to Make and Break Secret Codes

Name _____ **Date** _____

(Text page 163)

6. a. One attempt at a formula to generate prime numbers is $n^2 - n + 41$. Evaluate the formula for $n = 0, 1, 2, 3, 4, 5$. Do you always get a prime?

   _____

   _____

   _____

   _____

   _____

   _____

   b. **Challenge.** Find an n less than 50 for which the formula in 6a does not generate a prime.

   _____

   _____

7. Look back at your list of primes. Find all pairs of twin primes between 1 and 100.

   _____

   _____

8. Find the Mersenne numbers for $n = 5, 6, 7,$ and 11. Which of these are prime?

   _____

   _____

   _____

   _____

© 2006 A K Peters, Ltd., Wellesley, MA

(Text page 163)

9. Find at least three Sophie Germaine primes other than 2, 3, and 5.

    _____

    _____

    _____

    _____

10. **Challenge.** Find a large prime number. (You decide whether it is large enough to please you.) Explain how you chose the number and how you know it is prime.

    _____

    _____

    _____

    _____

    _____

    _____

    _____

    _____

    _____

    _____

    _____

    _____

© 2006 A K Peters, Ltd., Wellesley, MA

The Cryptoclub: Using Mathematics to Make and Break Secret Codes

(Text page 164)

11. a. Test the Goldbach Conjecture: Pick several even numbers greater than 2 and write each as the sum of two primes. (Don't use 1 in your sums, since 1 is not prime.)

_____

_____

_____

_____

b. Find a number that can be written as the sum of two primes in more than one way.

_____

_____

_____

© 2006 A K Peters, Ltd., Wellesley, MA

Continue to the next chapter.

© 2006 A K Peters, Ltd., Wellesley, MA

Name _____ Date _____

Chapter 17: Raising to Powers
(Text page 169)

1. Compute the following. Reduce before your numbers get too large.

   a. $482^4 \mod 1000$

   b. $357^5 \mod 1000$

   c. $993^5 \mod 1000$

   d. $888^6 \mod 1000$

© 2006 A K Peters, Ltd., Wellesley, MA

The Cryptoclub: Using Mathematics to Make and Break Secret Codes

(Text page 170)

2. a. How many multiplications would it take to compute 18^{32} mod 55 using the method of repeated squaring?

    b. How many multiplications would it take to compute 18^{32} mod 55 by multiplying 18 by itself over and over?

    c. Compute 18^{32} mod 55 using the method from 2a or 2b that uses the fewest multiplications. (You can reuse calculations from this chapter.)

© 2006 A K Peters, Ltd., Wellesley, MA

The Cryptoclub: Using Mathematics to Make and Break Secret Codes

(Text page 170)

3. Use the method of repeated squaring to compute each number.

   a. 6^8 mod 26

   b. 3^8 mod 5

© 2006 A K Peters, Ltd., Wellesley, MA

The Cryptoclub: Using Mathematics to Make and Break Secret Codes

(Text page 170)

3.   c. 9^{16} mod 11

d. 4^{16} mod 9

© 2006 A K Peters, Ltd., Wellesley, MA

The Cryptoclub: Using Mathematics to Make and Break Secret Codes

© 2006 A K Peters, Ltd., Wellesley, MA

Name _____ Date_____

(Text page 171)

4. Use some of the powers already computed in the text to find each value.

    a. 18^6 mod 55

    b. 18^{12} mod 55

    c. 18^{20} mod 55

The Cryptoclub: Using Mathematics to Make and Break Secret Codes

(Text page 171)

5. a. Make a list of the values 9^n mod 55 for $n = 1, 2, 4, 8,$ and 16. Reduce each expression.

    b. Combine your answers from 5a to compute 9^{11} mod 55.

    c. Combine your answers from 5a to compute 9^{24} mod 55.

© 2006 A K Peters, Ltd., Wellesley, MA

The Cryptoclub: Using Mathematics to Make and Break Secret Codes

Name _____ Date_____

© 2006 A K Peters, Ltd., Wellesley, MA

The Cryptoclub: Using Mathematics to Make and Break Secret Codes

(Text page 171)

6. a. Make a list of the values 7^n mod 31, for n = 1, 2, 4, 8, and 16. Reduce each expression.

b. Combine your answers from 6a to compute 7^{18} mod 31.

c. Combine your answers from 6a to compute 7^{28} mod 31.

Continue to the next chapter.

© 2006 A K Peters, Ltd., Wellesley, MA

Chapter 18: The RSA Cryptosystem
(Text pages 178 and 180)

1. Use Tim's RSA public encryption key (55, 7) to encrypt the word **fig**.
   (First change the letters to numbers using **a** = 0, **b** = 1, **c** = 2, etc.)

Answer: _____

Return to Text

2. Review: Show that 4^{23} mod 55 = 9.

© 2006 A K Peters, Ltd., Wellesley, MA

The Cryptoclub: Using Mathematics to Make and Break Secret Codes

(Text page 180)

3. Dan encrypted a word with Tim's encryption key $(n, e) = (55, 7)$. He got the numbers 4, 0, 8. Use Tim's decryption key $d = 23$ to decrypt these numbers and get back Dan's word. (Hint: You can use your result from Problem 2.)

© 2006 A K Peters, Ltd., Wellesley, MA

Answer: _____

The Cryptoclub: Using Mathematics to Make and Break Secret Codes

Name _____ Date_____

Chapter 19: Revisiting Inverses in Modular Arithmetic
(Text page 186)

1. For each of the following, determine whether the inverse exists in the given modulus. If it exists, use either Jenny's method or Evie's method to find it.

a. 10 (mod 13)

Answer: _____

b. 10 (mod 15)

Answer: _____

c. 7 (mod 21)

Answer: _____

© 2006 A K Peters, Ltd., Wellesley, MA

The Cryptoclub: Using Mathematics to Make and Break Secret Codes

(Text page 186)

1.   d. 7 (mod 18)

Answer: _____

e. 11 (mod 24)

Answer: _____

f. 11 (mod 22)

Answer: _____

© 2006 A K Peters, Ltd., Wellesley, MA

(Text page 186)

2. Find the inverse of each of the following numbers in the given modulus.

    a. 11 (mod 180)

Answer: _____

    b. 9 (mod 100)

Answer: _____

    c. 7 (mod 150)

Answer: _____

© 2006 A K Peters, Ltd., Wellesley, MA

The Cryptoclub: Using Mathematics to Make and Break Secret Codes

Continue to the next chapter.

© 2006 A K Peters, Ltd., Wellesley, MA

© 2006 A K Peters, Ltd., Wellesley, MA

The Cryptoclub: Using Mathematics to Make and Break Secret Codes

Name _____ Date _____

Chapter 20: Sending RSA Messages
(Text pages 189 and 192)

Class Activity

Follow the instructions in the text to choose your own RSA key. Use your own paper to record your calculations. Then record your encryption key here.

RSA encryption: n = _____ e = _____ .

Record your decryption key d and your primes p and q in a secret place so you won't forget it. (If you put it here, other people can decrypt messages sent to you. But if you lose it even you won't be able to decrypt.)

Return to Text

1. Use Dan's keyword **CRYPTO** to decrypt his Vigenère message to Tim.

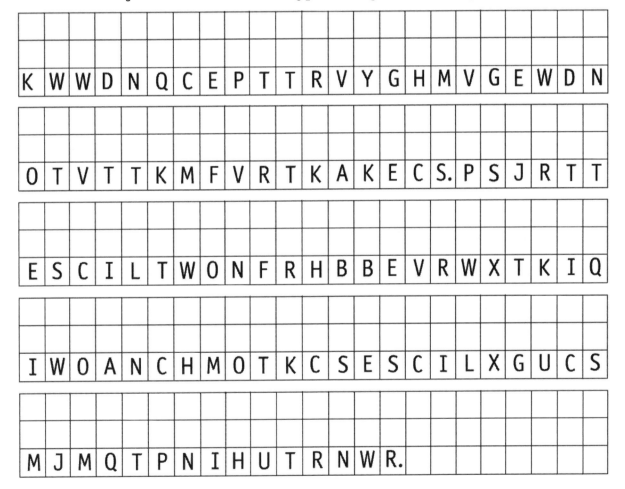

| K | W | W | D | N | Q | C | E | P | T | T | R | V | Y | G | H | M | V | G | E | W | D | N |

| O | T | V | T | T | K | M | F | V | R | T | K | A | K | E | C | S. | P | S | J | R | T | T |

| E | S | C | I | L | T | W | O | N | F | R | H | B | B | E | V | R | W | X | T | K | I | Q |

| I | W | O | A | N | C | H | M | O | T | K | C | S | E | S | C | I | L | X | G | U | C | S |

| M | J | M | Q | T | P | N | I | H | U | T | R | N | W | R. |

—Dan

(Text page 192)

2. Here is the reply Tim sent to Dan:

> Dan,
>
> Here is my reply. It is a Vigenère message. I used your RSA public key to encrypt my Vigenère keyword. This is what I got: 32, 209, 165, 140. You know what to do with it.
>
> ACXETSUMIVW.
>
> MCAGIVSUQKBHHCBGTTCXHVCR.
>
> —Tim

a. Dan's RSA decryption key is $d = 5$. Use it to find the keyword that Tim encrypted. (Tim used Dan's encryption key $(n,e) = (221, 77)$.) Show your work. Use the next page if you need more space.

© 2006 A K Peters, Ltd., Wellesley, MA

The Cryptoclub: Using Mathematics to Make and Break Secret Codes

Use this space to continue your work from 2a.

© 2006 A K Peters, Ltd., Wellesley, MA

(Text page 192)

2. b. Use the keyword you found in 2a to decrypt the Vigenère message Tim
sent to Dan.

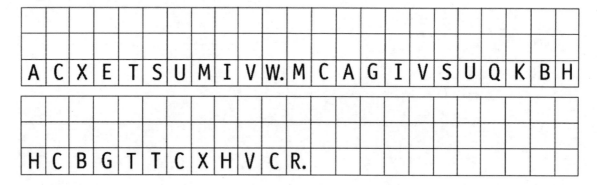

A	C	X	E	T	S	U	M	I	V	W.	M	C	A	G	I	V	S	U	Q	K	B	H

H	C	B	G	T	T	C	X	H	V	C	R.									

3. Follow the instructions in the text to combine RSA with the Vigenère
cipher and send an RSA message to someone. Use your own paper to write
your message and record your calculations.

© 2006 A K Peters, Ltd., Wellesley, MA

The Cryptoclub: Using Mathematics to Make and Break Secret Codes